João C. Setubal   Nalvo F. Almeida (Eds.)

Advances in Bioinformatics and Computational Biology

8th Brazilian Symposium on Bioinformatics, BSB 2013
Recife, Brazil, November 3-7, 2013
Proceedings

 Springer

Volume Editors

João C. Setubal
University of São Paulo, Institute of Chemistry
Avenida Prof. Lineu Prestes, 748 sala 911, 05508-000 São Paulo, SP, Brazil
E-mail: setubal@ig.usp.br

Nalvo F. Almeida
Federal University of Mato Grosso do Sul, School of Computing
Facom-UFMS, CP 549, 79070-900 Campo Grande, MS, Brazil
E-mail: nalvo@facom.ufms.br

ISSN 0302-9743   e-ISSN 1611-3349
ISBN 978-3-319-02623-7   e-ISBN 978-3-319-02624-4
DOI 10.1007/978-3-319-02624-4
Springer Cham Heidelberg New York Dordrecht London

Library of Congress Control Number: 2013949541

CR Subject Classification (1998): J.3, F.2, H.2, I.5, I.2, F.1, I.5

LNCS Sublibrary: SL 8 – Bioinformatics

Typesetting: Camera-ready by author, data conversion by Scientific Publishing Services, Chennai, India

Printed on acid-free paper

Springer is part of Springer Science+Business Media (www.springer.com)

Preface

This volume contains the papers selected for presentation at the 8th Brazilian Symposium on Bioinformatics (BSB 2013), held during November 3–6, 2013, in Recife, Brazil. BSB is an international conference that covers all aspects of bioinformatics and computational biology. This year the event was jointly organized by the special interest group in Computational Biology of the Brazilian Computer Society (SBC), which has been the organizer of BSB for the past several years, and by the Brazilian Association for Bioinformatics and Computational Biology (AB³C), which has been the organizer of another event, the X-Meeeting, also for the past several years. This year for the first time the two events were co-located.

As in previous editions, BSB 2013 had an international Program Committee (PC) of 46 members. After a rigorous review process by the PC, 18 papers were accepted to be orally presented at the event, and are printed in this volume. In addition to the technical presentations, BSB 2013 featured keynote talks from David Roos (University of Pennsylvania), Peter Stadler (University of Leipzig), and Martin Tompa (University of Washington). Peter Stadler graciously contributed an invited paper to these proceedings ("The Trouble with Long-Range Base Pairs in RNA Folding").

BSB 2013 was made possible by the dedication and work of many people and organizations. We would like to express our sincere thanks to all PC members, as well as to the external reviewers. Their names are listed in the pages that follow. We are also grateful to the local organizers and volunteers for their valuable help; the sponsors for making the event financially viable; Guilherme P. Telles and Peter Stadler for assisting in the preparation of the proceedings; and Springer for agreeing to publish this volume. Finally, we would like to thank all authors for their time and effort in submitting their work and the invited speakers for having accepted our invitation.

This year selected BSB 2013 papers were invited for submission in expanded format to a special issue of the *IEEE/ACM Transactions on Computational Biology and Bioinformatics*. We thank Ying Xu (Editor-in-Chief) and Dong Xu (Associate Editor-in-Chief) for this opportunity.

November 2013   João C. Setubal
                                                                       Nalvo F. Almeida

Organization

BSB 2013 was organized by Universidade Federal de Pernambuco.

Conference Chairs

Katia S. Guimarães	Universidade Federal de Pernambuco, Brazil
Paulo Gustavo S. Fonseca	Universidade Federal de Pernambuco, Brazil
Ana Benko Iseppon	Universidade Federal de Pernambuco, Brazil
Valdir Balbino	Universidade Federal de Pernambuco, Brazil

Program Chairs

João C. Setubal	Universidade de São Paulo, Brazil
Nalvo F. Almeida	Universidade Federal de Mato Grosso do Sul, Brazil

Steering Comittee

Guilherme P. Telles (President)	Universidade Estadual de Campinas, Brazil
Ivan G. Costa Filho	Universidade Federal de Pernambuco, Brazil
Luis Antonio Kowada	Universidade Federal Fluminense, Brazil
Marcelo de Macedo Brígido	Universidade de Brasília, Brazil
Marcilio C.P. de Souto	Université d'Orléans, France
Osmar Norberto de Souza	PUC Rio Grande do Sul, Brazil
Sérgio Vale Aguiar Campos	Universidade Federal de Minas Gerais, Brazil

Sponsors

Sociedade Brasileira de Computação (SBC)
Associação Brasileira de Bioinformática e Biologia Computacional (AB^{3}C)

Funding

Conselho Nacional de Desenvolvimento Científico e Tecnológico (CNPq)
Coordenação de Aperfeiçoamento de Pessoal de Nível Superior (CAPES)
Fundação de Amparo à Ciência e Tecnologia de Pernambuco (FACEPE)

Program Committee

Said S. Adi	Universidade Federal de Mato Grosso do Sul, Brazil
Nalvo F. Almeida	Universidade Federal de Mato Grosso do Sul, Brazil
Fernando Álvarez-Valín	Universidad de la Republica, Uruguay
Ana L.C. Bazzan	Universidade Federal do Rio Grande do Sul, Brazil
Ana M. Benko-Iseppon	Universidade Federal de Pernambuco, Brazil
Marília D.V. Braga	Inmetro, Brazil
Marcelo de Macedo Brígido	Universidade de Brasília, Brazil
Sergio Campos	Universidade Federal de Minas Gerais, Brazil
Alessandra Carbone	Pierre and Marie Curie University, France
Ivan Gesteira Costa	Universidade Federal de Pernambuco, Brazil
André C.P.L.F. de Carvalho	Universidade de São Paulo/São Carlos, Brazil
Marcilio de Souto	Université d'Orléans, France
Emanuel Maltempi de Souza	Universidade Federal do Paraná, Brazil
Osmar Norberto de Souza	PUC Rio Grande do Sul, Brazil
Luciano Digiampietri	Universidade de São Paulo, Brazil
Alan Durham	Universidade de São Paulo, Brazil
Carlos E. Ferreira	Universidade de São Paulo, Brazil
André Fujita	Universidade de São Paulo, Brazil
Katia S. Guimarães	Universidade Federal de Pernambuco, Brazil
Lenny Heath	Virginia Tech, USA
Dirk Husmeier	University of Glasgow, UK
Andre Kashiwabara	Universidade Tecnológica Federal do Paraná, Brazil
Ney Lemke	Universidade Est. Paulista Júlio Mesquita Filho, Brazil
Sergio Lifschitz	PUC Rio de Janeiro, Brazil
Ana Carolina Lorena	Universidade Federal do ABC, Brazil
Wellington Martins	Universidade Federal de Goiás, Brazil
Marta Mattoso	Universidade Federal do Rio de Janeiro, Brazil
Satoru Miyano	University of Tokyo, Japan
Houtan Noushmehr	Universidade de São Paulo, Brazil
Guilherme Oliveira	Fundação Oswaldo Cruz, Minas Gerais, Brazil
Alexey Onufriev	Virginia Tech, USA
Miguel Ortega	Universidade Federal de Minas Gerais, Brazil
Sergio Pantano	Instituto Pasteur de Montevideo, Uruguay
Alexandre Rossi Paschoal	Universidade Tecnológica Federal do Paraná, Brazil
Duncan Ruiz	PUC Rio Grande do Sul, Brazil
Marie-France Sagot	Inria, France
Alexander Schliep	Rutgers University, USA

João Setubal	Universidade de São Paulo, Brazil
Peter F. Stadler	University of Leipzig, Germany
Jens Stoye	University of Bielefeld, Germany
Guilherme P. Telles	Universidade Estadual de Campinas, Brazil
Thiago Venancio	Universidade Federal Fluminense, Brazil
Maria Emilia T. Walter	Universidade de Brasília, Brazil

Additional Reviewers

Poly Hannah Da Silva	Universidade Federal Fluminense, Brazil
Marcelo Henriques de Carvalho	Universidade Federal de Mato Grosso do Sul, Brazil
Zanoni Dias	Universidade Estadual de Campinas, Brazil
Marcio Dorn	Universidade Federal do Rio Grande do Sul, Brazil
Alessandra Faria-Campos	Universidade Federal de Minas Gerais, Brazil
Sonja Hänzelmann	Universitat Pompeu Fabra, Spain
Tobias Jakobi	University of Bielefeld, Germany
Pablo Jaskowiak	Universidade de São Paulo/São Carlos, Brazil
Thiago Lipinski Paes	PUC Rio Grande do Sul, Brazil
Fábio Viduani Martinez	Universidade Federal de Mato Grosso do Sul, Brazil
Mariana Mendoza	Universidade Federal do Rio Grande do Sul, Brazil
Kary Ocaña	Universidade Federal do Rio de Janeiro, Brazil
Luiz Otávio Murta Jr	Universidade de São Paulo/Ribeirão Preto, Brazil
Gethin Norman	University of Glasgow, UK
Yuri Pirola	University of Milano-Bicocca, Italy
Tainá Raiol	Universidade de Brasília, Brazil
Eric Tannier	Inria, France
Ian Thorpe	University of Maryland, USA

Table of Contents

The Trouble with Long-Range Base Pairs in RNA Folding 1
  Fabian Amman, Stephan H. Bernhart, Gero Doose, Ivo L. Hofacker,
  Jing Qin, Peter F. Stadler, and Sebastian Will

Roles of RORα on Transcriptional Expressions in the Mammalian
Circadian Regulatory System 12
  Hiroshi Matsuno and Makoto Akashi

HybHap: A Fast and Accurate Hybrid Approach for Haplotype
Inference on Large Datasets 24
  Rogério S. Rosa and Katia S. Guimarães

Restricted DCJ-Indel Model Revisited 36
  Marília D.V. Braga and Jens Stoye

MSA-GPU: Exact Multiple Sequence Alignment Using GPU 47
  Daniel Sundfeld and Alba C.M.A. de Melo

MultiSETTER - Multiple RNA Structure Similarity Algorithm 59
  David Hoksza, Peter Szépe, and Daniel Svozil

Assessing the Accuracy of the SIRAH Force Field to Model DNA at
Coarse Grain Level 71
  Pablo D. Dans, Leonardo Darré, Matías R. Machado, Ari Zeida,
  Astrid F. Brandner, and Sergio Pantano

How to Multiply Dynamic Programming Algorithms 82
  Christian Höner zu Siederdissen, Ivo L. Hofacker, and
  Peter F. Stadler

Influence of Scaffold Stability and Electrostatics on Top7-Based
Engineered Helical HIV-1 Epitopes 94
  Isabelle F.T. Viana, Rafael Dhalia, Marco A. Krieger,
  Ernesto T.A. Marques, and Roberto D. Lins

Random Forest and Gene Networks for Association of SNPs to
Alzheimer's Disease 104
  Gilderlanio S. Araújo, Manuela R.B. Souza,
  João Ricardo M. Oliveira, Ivan G. Costa, and
  for the Alzheimer's Disease Neuroimaging Initiative

Multilayer Cluster Heat Map Visualizing Biological Tensor Data   116
    Atsushi Niida, Georg Tremmel, Seiya Imoto, and Satoru Miyano

On the 1.375-Approximation Algorithm for Sorting by Transpositions
in $O(n \log n)$ Time   126
    Luís Felipe I. Cunha, Luis Antonio B. Kowada,
    Rodrigo de A. Hausen, and Celina M.H. de Figueiredo

ncRNA-Agents: A Multiagent System for Non-coding RNA
Annotation.   136
    Wosley Arruda, Célia G. Ralha, Tainá Raiol, Marcelo M. Brígido,
    Maria Emília M.T. Walter, and Peter F. Stadler

Inference of Genetic Regulatory Networks Using an Estimation of
Distribution Algorithm   148
    Thyago Salvá, Leonardo R. Emmendorfer, and Adriano V. Werhli

A Network-Based Meta-analysis Strategy for the Selection of Potential
Gene Modules in Type 2 Diabetes.   160
    Ronnie Alves, Marcus Mendes, and Diego Bonnato

RAIDER: Rapid Ab Initio Detection of Elementary Repeats.   170
    Nathaniel Figueroa, Xiaolin Liu, Jiajun Wang, and John Karro

A Probabilistic Model Checking Analysis of the Potassium Reactions
with the Palytoxin and Na^+/K^+-ATPase Complex   181
    Fernando Braz, João Amaral, Bruno Ferreira, Jader Cruz,
    Alessandra Faria-Campos, and Sérgio Campos

False Discovery Rate for Homology Searches.   194
    Hyrum D. Carroll, Alex C. Williams, Anthony G. Davis, and
    John L. Spouge

A Pipeline to Characterize Virulence Factors in *Mycobacterium*
Massiliense Genome.   202
    Guilherme Menegói, Tainá Raiol, João Victor de Araújo Oliveira,
    Edans Flávius de Oliveira Sandes,
    Alba Cristina Magalhães Alves de Melo, Andréa Queiroz Maranhão,
    Ildinete Silva-Pereira, Anamélia Lorenzetti Bocca,
    Ana Paula Junqueira-Kipnis, Maria Emília M.T. Walter,
    André Kipnis, and Marcelo de Macedo Brígido

Author Index   215

The Trouble with Long-Range Base Pairs in RNA Folding[*]

Fabian Amman[1], Stephan H. Bernhart[1], Gero Doose[1,2], Ivo L. Hofacker[3,5,8],
Jing Qin[1,4], Peter F. Stadler[1−7], and Sebastian Will[1]

[1] Dept. Computer Science, and Interdisciplinary Center for Bioinformatics,
Univ. Leipzig, Härtelstr. 16-18, Leipzig, Germany
[2] LIFE, Leipzig Research Center for Civilization Diseases, University Leipzig,
Philipp-Rosenthal-Strasse 27, 04107 Leipzig, Germany
[3] Dept. Theoretical Chemistry, Univ. Vienna, Währingerstr. 17, Wien, Austria
[4] MPI Mathematics in the Sciences, Inselstr. 22, Leipzig, Germany
[5] RTH, Univ. Copenhagen, Grønnegårdsvej 3, Frederiksberg C, Denmark
[6] FHI Cell Therapy and Immunology, Perlickstr. 1, Leipzig, Germany
[7] Santa Fe Institute, 1399 Hyde Park Rd., Santa Fe, USA
[8] Bioinformatics and Computational Biology research group, University of Vienna,
1090 Währingerstraße 17, Vienna, Austria

Abstract. RNA prediction has long been struggling with long-range base pairs since prediction accuracy decreases with base pair span. We analyze here the empirical distribution of base pair spans in large collection of experimentally known RNA structures. Surprisingly, we find that long-range base pairs are overrepresented in these data. In particular, there is no evidence that long-range base pairs are systematically overpredicted relative to short-range interactions in thermodynamic predictions. This casts doubt on a recent suggestion that kinetic effects are the cause of length-dependent decrease of predictability. Instead of a modification of the energy model we advocate a modification of the expected accuracy model for RNA secondary structures. We demonstrate that the inclusion of a span-dependent penalty leads to improved maximum expected accuracy structure predictions compared to both the standard MEA model and a modified folding algorithm with an energy penalty function. The prevalence of long-range base pairs provide further evidence that RNA structures in general do not have the so-called polymer zeta property. This has consequences for the asymptotic performance for a large class of sparsified RNA folding algorithms.

Keywords: RNA folding, long-range base pair, prediction accuracy, polymer zeta property.

1 Introduction

Despite the many successful applications of thermodynamics-based RNA secondary predictions it has remained a challenging problem to predict long-range

---

[*] The Students of the Bioinformatics II Lab Class 2013.

J.C. Setubal and N.F. Almeida (Eds.): BSB 2013, LNBI 8213, pp. 1–11, 2013.
© Springer International Publishing Switzerland 2013

base pairs with high accuracy. This deficit in the thermodynamic model has been known for a long time, see e.g. [1]. Despite recent efforts, however, we still lack a convincing solution to this problem.

Several authors have devised modified algorithms that treat a base pair (p, q) in dependence of its span $\ell = q - p + 1$, i.e., the number of nucleotides between its end points p and q measured along the backbone. In the simplest case, one strictly disallows the formation of long-range base pairs by limiting base pairs to a maximum span L. This has the convenient side effect that computation complexity of the folding algorithms drops from $O(N^3)$ to $O(NL^2)$. Several implementations of span-restricted folding have been published, including RNALfold [2] and its partition function version RNAplfold [3], as well as Rfold [4] and Raccess [5]. Thorough benchmarking [6] suggests an optimal span of $L \approx 150$ as a "reasonable balance between maximizing the number of accurately predicted base pairs, while minimizing effects of incorrect long-range predictions." Instead of applying a hard cutoff, CoFold [7] discounts long-range base pairs by a span-dependent penalty factor. This additional term is motivated as a net effect of kinetic folding: since local structures can form rapidly already during the course of transcription, many bases are kinetically trapped in local structures and hence are later-on not available for the formation of long-range base pairing.

Discouraging long-range base pairs in computational approaches, however, appears to be at odds with a wide variety of reports of functionally important long-range structures. The first well-described motifs of this type are the panhandle structures that are abundant throughout RNA virus genomes, see e.g. [8]. Furthermore, recent SHAPE-seq data provide direct evidence for functional long-distance interactions e.g. in the tombusvirus tomato bushy stunt virus (TBSV) [9]. Long-range base pairing is also involved in modulation of alternative splicing in a wide variety of mRNAs [10, 11].

A series of theoretical studies furthermore emphasizes that long-range pairing is an important, generic feature of RNA secondary structures. In particular, the 3' and 5' ends are typically close to each other [12–15]. Measured in terms of the natural graph distance on the secondary structures, the expected end-to-end distance remains asymptotically constant. This, at face value counter-intuitive mathematical result, has recently stimulated a more detailed investigation in the distribution of graph distances between arbitrary positions in an ensemble of RNA structures [16]. Here, we take a complementary point of view and ask for the distribution of base pairs as a function of the span L and the RNA sequence length n.

The (low) abundance of long-range base pairs is essential for sparse RNA folding algorithms [17]. For many of these approaches, an asymptotic performance gain was claimed under a specific assumption about the rapid drop of base pair probability with increasing span; this assumption is known as polymer-zeta property. Our results hint at the invalidity of this assumption.

Finally, we study the CoFold-like idea of penalizing long-range base pairs in more generality. For this purpose we suggest a novel maximum expected accuracy prediction approach. Strong sparsification of this approach allows for fast exploration of the prediction accuracy landscape of the parameter space.

2   Empirical Distribution of Base Pair Spans

Throughout this contribution we predict only pseudoknot-free RNA secondary structures, i.e., each two base pairs (i, j) and (k, l) satisfy that $i < k < j$ implies $i < k < l < j$ (as well as $j \neq k$ and $i = k \leftrightarrow j = l$.) We first consider the empirical evidence regarding the abundance of long-range base pairs. RNAstrand 2.0 [18] is a database of high quality structures compiled from several sources such as Rfam, the RCSB Protein Data Bank and some specialized databases such as the tmRNA database[19], SRP database[20] and RNaseP database[21]. Since we were interested in long-range interactions, we removed RNAs with less than 200 nucleotides. Furthermore, we discarded any molecule with more than two consecutive unknown nucleotides (Ns) or gaps. These pruning steps left us with our *data set* of 2010 structures.

For each of the data base entries, we predicted structures using CoFold at default parameter settings. As a control, we furthermore shuffled the entire data set preserving dinucleotide frequencies (using a reimplementation of [22]) and predicted structures with CoFold again. Then, we counted the base pairs of specific spans in the data set structures, the CoFold predictions from the data set sequences, and the CoFold predictions from dinucleotide shuffled data. The resulting empirical distributions are shown in Figure 1A.

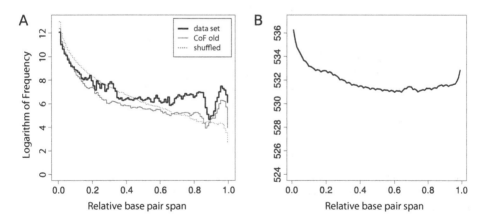

Fig. 1. A Empirical base pair span distribution. We plot base pair spans relative to the sequence lengths *vs.* the logarithm of their absolute frequency in a collection of the 2010 data set structures taken from the RNAstrand database (black). For comparison we show the span distribution of structure predictions with CoFold for the same sequences (blue) and for a dinucleotide shuffle control data set (dotted red). **B** Total number of canonical base pairs in the ensembles of all database sequences as a function of their relative base pair spans ℓ/n.

Surprisingly, both the empirical distribution and the `CoFold` predictions for the same sequences shows a pronounced over-representation for long-range base pairs compared to the randomized control. This enrichment is most likely the effect of natural selection and strongly suggests that long-range base pairs play an important role in at least a subset of RNA secondary structures included in the `RNAstrand` database. It cannot be explained by the known panhandle structures of viral RNAs since no complete viral genomes are contained in this resource. Fig. 1A also shows that the computational discouragement of long-range pairs leads to a systematic underestimation of long-range pairs, which is most pronounced around $\ell \approx 0.75n$.

To investigate the over-representation of long-range base pairs further, we decided to look at the structural ensembles of the selected `RNAstrand` sequences. Based on an adaption of Nussinov's algorithm, we counted the base pairs of span ℓ over all sequences in all of their secondary structures allowing only canonical base pairs with minimal loop size $m = 3$. The result, shown in Figure 1B, aids the interpretation of the empirical distributions of Figure 1A. As in the data set distribution, the number of base pairs does rise slightly starting at a relative span of 0.6, and has a peak at the end to end pairs. However, the dent at a relative base pair span of 0.9 is not present in the distribution of Fig. 1B.

3   Theoretical Distribution of Base Pair Spans

To understand the qualitative features of the span distribution of Figures 1A and B better, we consider long-range base pairs from a combinatorial perspective. More precisely, we ask, how prevalent long-range base pairs are already in the absence of energetic considerations.

To this end, consider a secondary structure S and denote by $\nu_S(\ell)$ the total number of base pairs in S with span ℓ. Similarly, let ν_S be the number of base pairs in S. We are interested in the distribution $P_n(\ell)$ of base pairs with span ℓ in the structure ensemble; here, the ensemble consists of all secondary structures over sequences of length n with minimal loop size m, i.e. each base pair (i, j) satisfies $j - i > m$. Setting $C_n(\ell) = \sum_S \nu_S(\ell)$ and $C_n = \sum_S \nu_S = \sum_\ell C_n(\ell)$ we obtain

$$P_n(\ell) = C_n(\ell)/C_n. \tag{1}$$

Let N_n denote the number of secondary structures in the ensemble, the expected number of base pairs with span ℓ in a secondary structure can be calculated from $P_n(\ell)$ as

$$e_n(\ell) = \frac{C_n(\ell)}{N_n} = \frac{C_n}{N_n} \cdot P_n(\ell). \tag{2}$$

In the absence of sequence constraints and influences of energy parameters on base pairing, the ensemble of secondary structures for an RNA sequence is identified with the set of noncrossing partial matchings on n vertices. Their number N_n of distinct structures can be obtained recursively as

$$N_n = N_{n-1} + \sum_{k=m+2}^{n} N_{k-2} N_{n-k} \tag{3}$$

with initial conditions $N_0 = N_1 = \cdots = N_m = N_{m+1} = 1$, see e.g. [23]. A similar decomposition can be used to compute $C_n(\ell)$: First, we observe that $C_n(\ell) = 0$ for $\ell < m + 2$, where m is the minimal number of unpaired bases in a hairpin loop. Furthermore, $C_n(\ell) = 0$ for $\ell > n$. Base pairs and their span remain unchanged when an unpaired base is appended. For structures with base pair $(1, k)$ we argue as follows: If $k < \ell + 2$, only the outside part can contain a pair of span ℓ. For a given k, there are N_{k-2} way to combine the internal part with the external one, on which we have a total of $C_{n-k}(\ell)$ base pairs of span ℓ. Similarly, if $k > n - \ell$, only the part inside the base pair may contain base pairs of span ℓ. For intermediate values of k, except for $k = \ell$, both combinations contribute. In the special case $k = \ell$ we have in addition $N_{\ell-2} N_{n-\ell}$ structures with the base pair $(1, \ell)$. In summary, we thus obtain the recursion

$$C_n(\ell) = C_{n-1}(\ell) + N_{\ell-2} N_{n-\ell} + \sum_{k=\ell+2}^{n} C_{k-2}(\ell) N_{n-k} + \sum_{k=m+2}^{n-\ell} N_{k-2} C_{n-k}(\ell). \tag{4}$$

To investigate the asymptotic behavior of the $C_n(\ell)$, and hence to better understand the distribution $P_n(\ell)$ for large n, we consider the bivariate generating function $\mathbf{C}(z, u) = \sum_n \sum_\ell C_n(\ell) z^n u^\ell$.

Multiply both sides of Eq. (4) with $z^n u^\ell$ and then sum over all possible n and ℓ, introducing the generating function $\mathbf{N}(z) = \sum_n N_n z^n$ for the N_n, and solving for $\mathbf{C}(z, u)$, we obtain

$$\mathbf{C}(z, u) = \frac{z^2 u^2 \cdot \mathbf{N}(z) \cdot (\mathbf{N}(zu) - \phi_m(zu))}{1 - z - 2z^2 \cdot \mathbf{N}(z) + z^2 \phi_m(z)}. \tag{5}$$

Here, $\phi_0(z) = 0$ and $\phi_m(z) = \sum_{k=0}^{m-1} N_k z^k$ for $m \geq 1$. The asymptotic behavior of the coefficients depends on the singularities of $C(z, u)$. We observe a discontinuity for $u = 1$ that is reminiscent of what is known as phase-transition phenomena in statistical physics. For $|u| \leq 1$; the dominant singularity of $\mathbf{C}(z, u)$ ρ; it is of the same type as the dominant singularity ρ_f of $f(z) = \frac{z^2 u^2 \cdot \mathbf{N}(z)}{1 - z - 2z^2 \cdot \mathbf{N}(z) + z^2 \phi_m(z)}$. For $|u| > 1$, however, the singularity is located at $\rho(u) = \min\{\rho_f, \rho_{\mathbf{N}}/u\}$, where $\rho_{\mathbf{N}}$ is the dominant singularity of $\mathbf{N}(z)$. This considerably complicates the computation of the limit distribution and will be investigated in detail elsewhere.

Here we consider only the simplest case $m = 0$ with all bases paired, i.e., the ensemble of non-crossing matchings, for which explicit expressions can be obtained. Using $\mathbf{N}(z) = (1 - \sqrt{1 - 4z^2})/(2z^2)$, Eq. 5 simplifies to $\mathbf{C}(z, u) = \frac{(1-\sqrt{1-4z^2})(1-\sqrt{1-4z^2 u^2})}{4z^2 \sqrt{1-4z^2}}$, from which the coefficients can be obtained explicitly. Making use of the usual approximation of central binomial coefficients, we arrive at

$$P_n(\ell) \sim \frac{n+2}{\sqrt{\pi}\ell} \sqrt{\frac{2}{(n - \ell + 2)n(\ell - 2)}}. \tag{6}$$

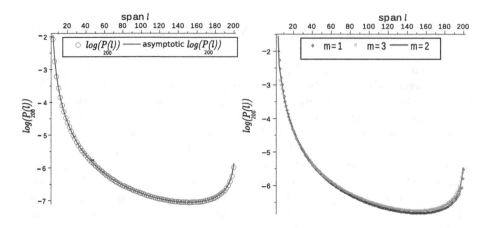

Fig. 2. Logarithmic scale of the probability distribution $P_n(\ell)$. **(Left)** Comparison of $P_{200}(\ell)$ and its asymptotic approximation for $m = 0$ in the absence of unpaired bases. **(Right)** Qualitatively similar distributions are obtained for minimum hairpin sizes $m = 1, 2, 3$.

Note here, in the case of noncrossing matching ensemble, both n and ℓ are even numbers. According to Eq. 2, we derive the expected number of base pairs with span ℓ in a secondary structure as

$$e_n(\ell) = \frac{n}{2} P_n(\ell) \sim \frac{n+2}{2\sqrt{\pi \ell}} \sqrt{\frac{2n}{(n-\ell+2)(\ell-2)}}. \tag{7}$$

In particular we obtain $e_n(n) \sim 1/(2\sqrt{\pi})$, i.e., a constant probability for a base pair connecting the terminal bases, in line with previous combinatorial considerations.

As shown in Fig. 3, qualitatively similar distributions are obtained for minimum hairpin sizes $m = 0, 1, 2, 3$, which indicates that these distributions probably share the same type of limit distribution. When $n = 200$ for $m = 1, 2, 3$, the probability for observing base pairs with very long span scale within the "tail" around $[190, 200]$ is higher than those base pairs with span scale within $[180, 190]$. In other words, these probabilities are not simply exponentially decreased. As m increases, the probability to see long-range pairs only slightly decreased, which indicates that changes in the minimum hairpin sizes have very small effects on their limit distributions.

4   Span-dependent Scoring for Structure Prediction

The overabundance of long-range base pairs in the known secondary structures, Fig 1, indicates that one cannot simply disregard them altogether. The systematic underrepresentation of long-range base pairs in the CoFold [7] predictions, on the other hand, casts doubt on the assumption that long-range base pairs are discouraged by kinetic effects and hence can effectively be modeled by increasingly smaller stabilizing energy contributions as the span increases. It may, for a given nucleotide, be simply a consequence of the trivial fact that there are many possible alternative long-range pairings. Instead of introducing a long-range penalty into the energy model as in CoFold we therefore pursue a maximum expected accuracy (MEA) approach in which we discount the expected accuracy of a thermodynamically predicted base pair in a length-dependent manner.

The key idea of MEA approaches is to model an accuracy score $s(R)$ for every given structure R and to compute the structure that maximized this score. The base pairing probabilities p_{ij} as computed by [24] provide the natural starting point for computing $s(R)$. MEA has proved to be a very accurate method to predict "best" secondary structures from the base pair probabilities in Boltzmann-distributed RNA ensembles [25]. In order to account for the reported problems with predicting long-range base pairs we introduce a span-dependent penalty $\pi(\ell)$ into the definition of the accuracy score $s(R)$ for a given secondary structure R:

$$s(R) = \sum_{(i,j)\in R} 2\gamma\pi(j-i)p_{ij} + \sum_{i\notin R} p^u(i), \tag{8}$$

$p^u(i) = 1 - \sum_j p_{(i,j)}$ denotes the probability that i is unpaired given P and $i \notin R$ refers to nucleotides that are unpaired in R.

The standard MEA ansatz for RNA folding (e.g., [26]) is recovered by $\pi(\ell) = 1$. The MEA score is optimized using the dynamic programming recursion

$$M(i,j) = \max \begin{cases} M(i,j-1) + p^u(j) \\ \max_{i<k<j} M(i,k-1) + M(k,j) \\ M(i+1,j-1) + 2\gamma\pi(j-i)p_{ij} \end{cases} \tag{9}$$

for $i \geq j$ and $M(i,j) = 0$, otherwise. Compared to the thermodynamic energy minimization, the MEA algorithm is much simpler since the complications incurred by loop-based energy model are already included in the computation of the base pairing probability matrix $P = (p_{ij})$ by McCaskill's algorithm [24] using here the implementation provided by the ViennaRNA package [26].

We consider here the general sigmoidal function

$$\pi(\ell) = \delta + \frac{1-\delta}{1 + \exp(\alpha(\ell - \beta)}$$

depending on the three parameters α, β, and δ. For $\delta = 1$, the model gracefully reduces to standard MEA. This model is substantially more general than the simple decay used in CoFold.

The simple structure of the MEA recursion, Eq. (9) allowed us to implement the highly optimized variant with OCT-STEP sparsification as described in [17] for the variant without span-dependent penalty. This drastically reduces computational cost, in particular for the LSU rRNA sequences contained in the training

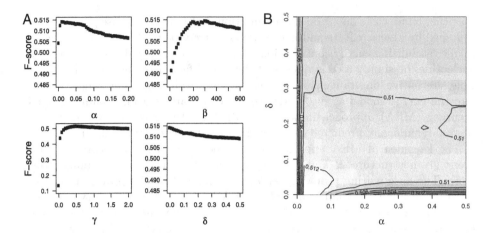

Fig. 3. Parameter estimation. **A** F-scores as a function of one parameter for the estimated optimum value of the other three parameters. **B** Contour plot of the F-score landscape as a function of (α, δ) for β and γ set to the trained parameters, generated from 25×25 equidistantly spaced values of (α, δ).

set with $n \approx 3500$, and allowed the systematic exploration of parameter space for the penalty function.

We evaluate the dependence of the prediction accuracy on the four parameter parameters of the penalty function by calculating the average F-score [27] of predictions to reference structures from the RNA strand database [18]. From the data set of Section 2, we select a training set and a disjoint control set of 100 sample instances each. Figure 3A illustrates our learning strategy: we iteratively scan over values of one of the four parameters while holding the other three parameters at previous optima. After three iterations, computing 20 values in each scan (performed in the order γ-β-α-δ), we find optimal parameters $\alpha = 0.012$, $\beta = 316$, $\gamma = 0.505$, $\delta = 0.003$, which yield an F-score of about 0.514. Figure 3B shows the two-dimensional landscape of F-scores as a function of the parameters α and δ, where the β and γ are set to their previously estimated optima.

The prediction and F-score evaluation is implemented in C++, where the optimization itself is performed by R interfacing to C++. The software package MEA is freely available at http://www.bioinf.uni-leipzig.de/Software/mea.

Using the control set, we compare the performance of the novel MEA approach at the learned parameters to RNAfold minimum free energy (MFE) prediction, MFE predictions with hard span cutoffs at maximal spans $L = 100$, $L = 150$ and $L = 200$, CoFold predictions (with default settings) and standard MEA predictions from the McCaskill ensembles, see Table 1.

Not unexpectedly, hard cut-off values L for the span do not perform well on the RNAstrand data set of long RNAs (even more since long-range base pairs are overrepresented.) The energy penalties of CoFold also cannot properly capture long-range effects in this data set. We observe, in fact, reduction of the F-measure

Table 1. Comparison of prediction quality. Average F-scores for training and control data sets by various prediction methods.

data set	MFE					MEA	
	RNAfold	CoFold	$L = 100$	$L = 150$	$L = 200$	standard	novel
training	0.467	0.449	0.352	0.386	0.393	0.497	0.514
control	0.473	0.418	0.332	0.363	0.385	0.486	0.496

relative to standard RNAfold. The trained sigmoidal penalty function π provides a small advantage over the standard, span-independent MEA model, which in turns performs slightly better than the pure thermodynamic folding.

Although performance evaluations of folding procedures traditionally rely on Matthews correlation coefficient, we advocate the use of the F-measure. While the numbers are usually very well correlated, the F-measure does not require committing to a specific definition of false negatives, which in the case of RNA structures seems not obvious. Deviating from [28], we consider "false positive" all predicted non-reference base pairs, even if they are consistent with the reference. Consequently, the F-measures in Tab. 1 are systematically lower than the MCC values reported in [28].

5   Discussion

Long-range base pairs have long been known to cause difficulties for the prediction of RNA structures. An empirical analysis of the distribution of base pair spans in the experimentally determined structures compiled in the RNAstrand database shown that long-range base pairs are not depleted in biologically relevant structures. This casts doubt on the interpretation of [7] that the poor performance of thermodynamic folding algorithms in particular for long-range pairs is the consequence of kinetic effects that would frequently preclude the formation of long-range pairs, and hence would result in a systematic overprediction for large spans. We suggest that the decrease in the accuracy of pairs with large spans may rather derive from a larger number of possible alternatives with large ℓ and propose to include this effect in MEA framework rather than as a modification of the energy model. After training a span-dependent penalty function, we find a small but noticeable improvement in the prediction performance for the modified MEA approach relative to both undiscounted MEA and thermodynamic penalty functions.

The prevalence of long-range base pairs has also important consequence for a large class of sparsified algorithms that rely on the so-called "polymer-zeta property" to guarantee their performance gain [29]. The polymer-zeta property implies that the probability that the terminal bases of a folded RNA are paired decreases asymptotically with sequence length as n^{-c} for some $c > 1$. Our results, both for the combinatorics of long-range base pairs and for the empirical RNA structures in the RNAstrand database, strongly suggest that $c = 0$, i.e., the probability of terminal base pairs settles down at a constant value. Not only do biological relevant RNAs not have the polymer zeta property, but they favor long-range interactions. In particular, end-to-end base pairs have an asymptotically non-vanishing probability. This is consistent with an empirical analysis based on thermodynamic folding [30] as well as a detailed combinatorial analysis of certain sparsification approaches [31]. As consequence, sparsification schemes of the type employed by [29] can incur only a constant factor as performance gain.

The work reported here of course does not settle all the problems and questions associated with long-range base pairs. It represents only a first step of progress that has arisen in the context of a two week intensive Master-level computer lab course held at Univ. Leipzig in 2013. Many open questions remain. We still have no satisfactory explanation, either statistical or biochemical, for the decrease in prediction accuracy with span, and hence we do not know if the functional form the penalty function π is the best one, or even tenable from a theoretical point of view. It remains unexplored, furthermore, whether other features, such as local sequence composition, also have a significant impact on prediction accuracy.

Acknowledgments. The following students contributed to this research as part of the Bioinformatics II lab class 2013 "Bioinformatisches Praktikum des Moduls 10-202-2208 (Bioinformatik von RNA- und Proteinstrukturen)": Sarah Berkemer, Lisa Duchstein, Lieselotte Erber, Daniel Gerighausen, Sandra Gerstl, Benjamin Standfuß, and Tariq Yousef.

References

1. Doshi, K., Cannone, J., Cobaugh, C., Gutell, R.: Evaluation of the suitability of free-energy minimization using nearest-neighbor energy parameters for RNA secondary structure prediction. BMC Bioinformatics 5, 105 (2004)
2. Hofacker, I.L., Priwitzer, B., Stadler, P.F.: Prediction of locally stable RNA secondary structures for genome-wide surveys. Bioinformatics 20, 191–198 (2004)
3. Bernhart, S., Hofacker, I.L., Stadler, P.F.: Local RNA base pairing probabilities in large sequences. Bioinformatics 22, 614–615 (2006)
4. Kiryu, H., Kin, T., Asai, K.: Rfold: an exact algorithm for computing local base pairing probabilities. Bioinformatics 24, 367–373 (2008)
5. Kiryu, H., Terai, G., Imamura, O., Yoneyama, H., Suzuki, K., Asai, K.: A detailed investigation of accessibilities around target sites of siRNAs and miRNAs. Bioinformatics 27, 1788–1797 (2011)
6. Lange, S.J., Maticzka, D., Möhl, M., Gagnon, J.N., Brown, C.M., Backofen, R.: Global or local? Predicting secondary structure and accessibility in mRNAs. Nucleic Acids Res. 40, 5215–5226 (2012)
7. Proctor, J.R.P., Meyer, I.M.: CoFold: an RNA secondary structure prediction method that takes co-transcriptional folding into account. Nucleic Acids Res. 41, e102 (2013)
8. Romero-López, C., Berzal-Herranz, A.: A long-range RNA-RNA interaction between the 5' and 3' ends of the HCV genome. RNA 15, 1740–1752 (2009)
9. Wu, B., Grigull, J., Ore, M.O., Morin, S., White, K.A.: Global organization of a positive-strand RNA virus genome. PLoS Pathog. 9, e1003363 (2013)

10. Raker, V.A., Mironov, A.A., Gelfand, M.S., Pervouchine, D.D.: Modulation of alternative splicing by long-range RNA structures in Drosophila. Nucleic Acids Res. 37, 4533–4534 (2009)
11. Pervouchine, D.D., Khrameeva, E.E., Pichugina, M.Y., Nikolaienko, O.V., Gelfand, M.S., Rubtsov, P.M., Mironov, A.A.: Evidence for widespread association of mammalian splicing and conserved long-range RNA structures. RNA 18, 1–15 (2012)
12. Yoffe, A.M., Prinsen, P., Gelbart, W.M., Ben-Shaul, A.: The ends of a large RNA molecule are necessarily close. Nucl. Acids Res. 39, 292–299 (2011)
13. Fang, L.T.: The end-to-end distance of RNA as a randomly self-paired polymer. J. Theor. Biol. 280, 101–107 (2011)
14. Clote, P., Ponty, Y., Steyaert, J.M.: Expected distance between terminal nucleotides of RNA secondary structures. J. Math. Biol. 65, 581–599 (2012)
15. Han, H.S., Reidys, C.M.: The 5'-3' distance of RNA secondary structures. J. Comput. Biol. 19, 867–878 (2012)
16. Backofen, R., Fricke, M., Marz, M., Qin, J., Stadler, P.F.: Distribution of graph-distances in Boltzmann ensembles of RNA secondary structures. In: Darling, A., Stoye, J. (eds.) WABI 2013. LNCS, vol. 8126, pp. 112–125. Springer, Heidelberg (2013)
17. Backofen, R., Tsur, D., Zakov, S., Ziv-Ukelson, M.: Sparse RNA folding: Time and space efficient algorithms. J. Discr. Alg. 9, 12–31 (2011)
18. Andronescu, M., Bereg, V., Hoos, H.H., Condon, A.: RNA STRAND: the RNA secondary structure and statistical analysis database. BMC Bioinf. 9, 340 (2008)
19. Zwieb, C., Gorodkin, J., Knudsen, B., Burks, J., Wower, J.: tmrdb (tmrna database). Nucleic Acids Res. 31(1), 446–447 (2003)
20. Rosenblad, M.A., Larsen, N., Samuelsson, T., Zwieb, C.: Kinship in the SRP RNA family. RNA Biol. 6(5), 508–516 (2009)
21. Brown, J.: The ribonuclease p database. NAR 27(1) (1999)
22. Jiang, M., Anderson, J., Gillespie, J., Mayne, M.: ushuffle: a useful tool for shuffling biological sequences while preserving the k-let counts. BMC Bioinformatics 9(1), 192 (2008)
23. Waterman, M.S.: Secondary structure of single-stranded nucleic acids. Adv. Math. Suppl. Studies 1, 167–212 (1978)
24. McCaskill, J.S.: The equilibrium partition function and base pair binding probabilities for RNA secondary structure. Biopolymers 29(6-7), 1105–1119 (1990)
25. Lu, Z., Gloor, J., Mathews, D.: Improved RNA secondary structure prediction by maximizing expected pair accuracy. RNA 15, 1805–1813 (2009)
26. Lorenz, R., Bernhart, S.H., Höner Zu Siederdissen, C., Tafer, H., Flamm, C., Stadler, P.F., Hofacker, I.L.: ViennaRNA Package 2.0. Algorithms Mol. Biol. 6, 26 (2011)
27. van Rijsbergen, C.J.: Information Retrieval. Butterworth (1979)
28. Gardner, P.P., Giegerich, R.: A comprehensive comparison of comparative RNA structure prediction approaches. BMC Bioinformatics 5, 140 (2004)
29. Wexler, Y., Zilberstein, C., Ziv-Ukelson, M.: A study of accessible motifs and RNA folding complexity. J. Comput. Biol. 14, 856–872
30. Dimitrieva, S., Bucher, P.: Practicality and time complexity of a sparsified RNA folding algorithm. J Bioinf. Comp. Biol. 10, 1241007 (2012)
31. Huang, F.W.D., Reidys, C.M.: On the combinatorics of sparsification. Alg. Mol. Biol. 7, 28 (2012)

Roles of RORα on Transcriptional Expressions in the Mammalian Circadian Regulatory System

Hiroshi Matsuno[1] and Makoto Akashi[2]

[1] Graduate School of Science and Engineering, Yamaguchi University
matsuno@sci.yamaguchi-u.ac.jp
[2] The Research Institute for Time Studies, Yamaguchi University
akashima@yamaguchi-u.ac.jp

Abstract. REV-ERBα and RORα are involved in the molecular regulatory system of mammalian circadian cycles, expressing opposite interactions on *Bmal1* expression, inhibition and activation, respectively. REV-ERBα has been thought to be the major regulator of gene expressions in phases, which is more than the role of RORα. This paper gives a contrary result to this, showing a prominent role of RORα in determining phase relations of the gene expression cycles. Computer simulations are conducted for the predictions of this RORα role, in addition, one of these predictions is supported by a biological experiment that shows combinatory effect of RORα and CRY on *Bmal1* transcription.

1 Introduction

Molecular mechanisms of the circadian rhythm for producing an endogenous oscillation with a period close to 24hr have been extensively investigated in many living organisms including bacteria, insects, and mammals. In the late 90s, the 4 genes *Per*, *Cry*, *Bmal1* and *Clock* have been mainly studied as central regulatory genes of mammalian circadian clock [1]. BMAL1/CLOCK, the complex of gene products of the *Bmal1* and *Clock* genes, is translocated to the nucleus and activates the transcriptions of the *Per* and *Cry* genes. PER/CRY, the complex of gene products of *Per* and *Cry* genes, is translocated to the nucleus and represses both the activations of the *Per* and *Cry* from the BMAL1/CLOCK. It is known that transcription cycles of the *Per* and *Cry* are nearly antiphase to that of the *Bmal1* in oscillation [2]. To understand the dynamics of these multilevel combination of genes, mRNAs, and proteins, many computational models have been suggested for the simulation of these complex behaviors [4].

In addition to these 4 genes and these products, REV-ERBα has been identified as the major regulator of cyclic *Bmal1* transcription [3], which represses the *Bmal1* transcription, while being regulated by PER/CRY and BMAL1/CLOCK in the same way as *Per* and *Cry*. Soon after the publication of this paper, several computational models including this REV-ERBα interaction were presented [5–8]. Especially, in the paper [7], incorporation of the REV-ERBα interaction to the existing model [9] was conducted, succeeding in reproducing the same

J.C. Setubal and N.F. Almeida (Eds.): BSB 2013, LNBI 8213, pp. 12–23, 2013.

simulation result as the experimental data [3] in *Bmal1* concentration in *Rev-Erbα*-knockout mice.

RORα[1] has been identified as a regulator that performs the contrary effect to REV-ERBα, that is, it activates the *Bmal1* transcription [10, 11]. Competitive effects of REV-ERBα and RORα on *Bmal1* transcription were examined in [12, 13]. A couple of computational models that include ROR interactions have been proposed. In the paper [14], although the importance of RORγ and REV-ERBα was clearly stated, contribution of these two proteins to the circadian genetic clock are not characterized well. A delay-differential equations model is constructed to realize accurate simulations of circadian cycles in both of amplitudes and phases, where REV-ERBα and RORγ interactions are appropriately incorporated [15]. The combinatory effect between ROR/REV-ERBα/BMAL1 loop and PER/CRY loop was investigated with simulations in [16], which demonstrates the role of ROR and REV-ERBα that forms a compensative loop to the PER/CRY loop.

Recently, the roles of ROR are extensively studied in the relations to other circadian gene expression patterns in different peripheral tissues [17] and in a pathological role in autoimmune disease [18]. A potential for the therapeutic target for breast cancer is suggested in [19]. Despite such increased attentions to RORα in the recent research activities, the importance of ROR as the main regulatory element for circadian cycles has not been well demonstrated yet due to, for example, the following reasons; no effect on other circadian genes in oscillation by the loss of ROR [10, 11]; advantage of REV-ERBα to ROR in the contribution to *Bmail1* in its rhythmic expressions and phase relations to other genes [13].

In this paper, a novel regulatory role of RORα on the phase regulation of *Bmal1* is predicted from a simulation with a hybrid functional Petri net model [20] of the genetic circadian system. A further prediction on RORα and CRY interaction is made from the results of a simulation of *Cry* knockout that shows an inconsistent *Bmal1* expression to a biological data. This prediction is supported by a biological experiment that shows combinatory effect of RORα and CRY on *Bmal1* transcription.

2   Basic Model of Circadian Gene Regulatory Network

Hybrid functional Petri net (HFPN) [20] is a modeling method for the simulation of biological reactions in a cell. We have constructed an HFPN model of basic circadian genetic control mechanism consisting of the five essential genes; *Per*, *Cry*, *Rev-Erb*, *Bmal*, and *Clock* [21]. In the following subsections, we present this HFPN model after giving brief description on the regulatory mechanism of these five genes.

---

[1] There are three subtypes RORα, RORβ, and RORγ. Since the effect of these three subtypes on *Bmal1* are similar, we use ROR to totally represent these three subtypes.

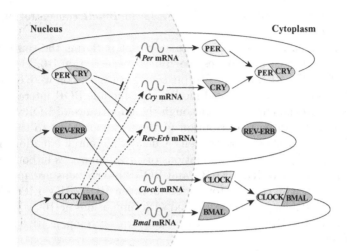

Fig. 1. Basic model of circadian gene regulatory system. *Per* and *Cry* are transcribed by CLOCK/BMAL, and translated into proteins that form heterodimers before returning to the cytoplasm. Products of *Clock* and *Bmal* heterodimerize to form the positive transcription factor for *Per*, *Cry*, and *Rev-Erb*; their effects are counteracted by PER/CRY. REV-ERB represses transcription of the *Bmal*.

2.1 Genetic Interactions of Five Basic Genes

A molecular clock reside within suprachiasmatic nucleus (SCN) cells. Each molecular circadian clock comprises a negative feedback loop of gene transcription and translation into the appropriate protein. The loop includes several genes and their protein products. In the case of mammals, 3 Period genes (*Per1*, *Per2* and *Per3*) and 2 Cryptochrome genes (*Cry1* and *Cry2*) comprise the negative limb, while the *Clock* and *Bmal1* (*Bmal*) genes constitute the positive limb of the feedback loop. In order to simplify the model and gain an insight into each interaction pathway, we examine two groups of genes: *Per1*, *Per2*, and *Per3* and *Cry1* and *Cry2*, collectively referred to as *Per* and *Cry*, respectively. *Rev-Erbα* (*Rev-Erb*) gene connects these positive and negative limbs. *Bmal* transcription is suppressed by REV-ERB, whose transcription is regulated by 2 interactions, activation by CLOCK/BMAL and repression by PER/CRY. Fig. 1 summarizes these interactions. Since *Per* and *Cry* and their products express the identical behaviors because of the symmetrical arrangements that exist in this model, we will only discuss the behaviors of *Cry* and its product in subsequent portions.

2.2 Hybrid Functional Petri Net Model

Fig. 2 shows an HFPN model of the basic circadian genetic control mechanism shown in Fig. 1, which was developed in Matsuno et al. [21]. HFPN allows us to model a system of biological reactions in a cell without utilizing

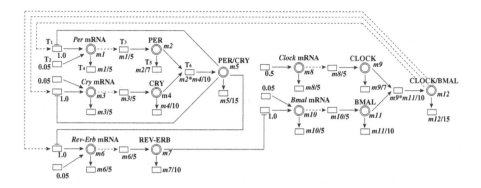

Fig. 2. HFPN model of the circadian genetic control mechanism shown in Fig. 1. A continuous place (doubled circle) holds a certain concentration of a gene product (mRNA or protein), and a formula representing the speed of a biological reaction is assigned at a continuous transition (rectangle). A continuous place and a continuous transition are connected by an arc that is chosen from a normal arc, test arc, or inhibitory arc depending on the biological relationship between two specified molecules.

any skills of mathematical descriptions and programming techniques. HFPN can manage 2 types of data, continuous and discrete, and consists of 3 types of elements, places, transitions, and arcs. However, discrete elements, namely, discrete places and discrete transitions, were not used in the HFPN model in this study.

A continuous place holds a real number such as the concentration of a substance, (e.g. mRNA or protein). A continuous transition is used to represent a biological reaction such as transcription and translation, and a formula representing the speed of a biological reaction is assigned at a continuous transition. There are 3 types arcs, normal, test, and inhibitory, which connect places and transitions. A normal arc is used to represent a flow of the substance, directed from the place to the transition or vice versa. A test or inhibitory arc is used to represent a condition and is only directed from a place to a transition. At the normal arcs from a place, test arcs, and inhibitory arcs, specified threshold values are assigned. The firing of the transition at the head of these arcs is controlled by the threshold value, namely, the normal arc from a place or the test arc (the inhibitory arc) can participate in activating (repressing) the transition at its head, so long as the content of the place at its tail is over the threshold value. For either test or inhibitory arcs, no amount is consumed from a place at its tail. The formal definition of an HFPN has been provided in the paper [20]. Simulations of HFPN models can be carried out with the software named "Cell Illustrator" [22].

3   A New Hypothesis: $Ror\alpha$ is the Major Regulator of Antiphasic Relationship between Cry and $Bmal1$

3.1   REV-ERBα is Not the Major Regulator of Antiphasic Relationship between Cry and $Bmal1$

Cry and $Bmal$ transcriptions are antiphasic [23]. In the summary of the paper [3], the following sentence occurs: "REV-ERBα constitutes a molecular link through which components of the negative limb drives antiphasic expression of components of the positive limb." That is, this paper made an assertion that REV-ERB is the major regulator that drives the antiphasic transcription cycles of $Bmal$ and Cry. However, our previous study [21] using the HFPN model of Fig. 2 produced a result that contradicts this assertion, namely, "antiphasic transcription cycles between $Bmal$ and Cry cannot be regulated by REV-ERBα alone." The summary of this discussion is given in Fig. 3.

3.2   An HFPN Model Incorporating Ror Interaction

Fig. 4 shows the circadian gene regulatory system that incorporates the RORα (ROR) interaction. We constructed an HFPN model of this system, which is illustrated in Fig. 5.

3.3   ROR Can Locate the $Bmal$ Peak at the Mid Point of Cry Peaks

In order to evaluate the effect of ROR on $Bmal$ transcription, we simulated this HFPN model in which the pale part in Fig. 5 that depicts the $Rev\text{-}Erb$ interaction was eliminated. After manipulating the parameters slightly, the $Bmal$ peak shifted to between (mid point) 2 successive Cry peaks, as shown in Fig. 6A. According to this model, $Bmal$ expression is controlled by only ROR. This means that an appropriate choice of the threshold value in the decreasing of ROR enables the shift in $Bmal$ to between (mid point) 2 successive Cry peaks. This implies that the antiphasic expression of the component of the negative limb (Cry) and that of the positive limb ($Bmal$) is not caused by REV-ERB but ROR.

4   Our Computer Simulation Suggests a Possible Novel Interaction between ROR and CRY

It is known that $Bmal$ is not expressed in Cry mutant mice [2]. However, the oscillation in $Bmal$ expression persisted in the HFPN simulation, in which the two arcs directed toward continuous place Cry mRNA are removed. (Fig. 6B). In order to resolve this inconsistency, we conducted an additional simulation with the HFPN model after incorporating the following hypothesis: "ROR interacts with CRY." Fig. 7 shows the HFPN model in which this hypothetical interaction has been incorporated. Fig. 8 provides the simulation results obtained from Cell

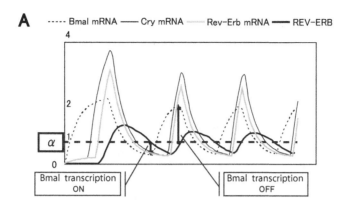

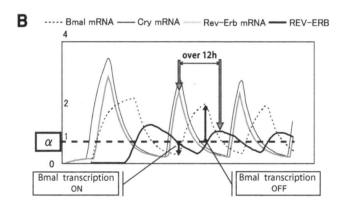

Fig. 3. (A) Simulation results of *Cry*, *Bmal*, and *Rev-Erb* transcriptions and REV-ERB translation behaviors from Cell Illustrator that were obtained for the HFPN model of Fig. 2. REV-ERB drives *Bmal* expression when the parameter α is assigned as the threshold. The transcription of *Bmal* is ON (OFF) at the point when the REV-ERB level crosses the threshold α downward (upward). In this figure, *Bmal* and *Cry* do not show an antiphasic relationship. **(B)** This figure illustrates the only condition under which the antiphasic relationship between *Bmal* and *Cry* is realized with REV-ERB regulation (not from simulation). Note that a *Bmal* peak can be located between 2 successive *Cry* peaks only when REV-ERB is on the rise. Considering the fact that the *Rev-Erb* peak is present at the same position as the *Cry* peak [25], we arrive at the unrealistic conclusion that REV-ERB requires over 12 h for its translation.

Illustrator. The results reveal that **(1)** the *Bmal* peak is present approximately between (mid point) successive *Cry* peaks and that **(2)** *Bmal* is not expressed when *Cry* expression is disrupted. Hence, the hypothetical path resolves the previous inconsistency.

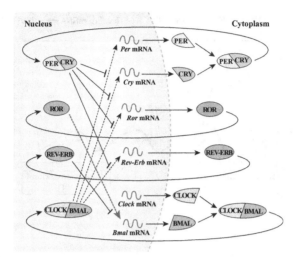

Fig. 4. Interaction among circadian genes after incorporating the *Ror* gene and its product into Fig. 1

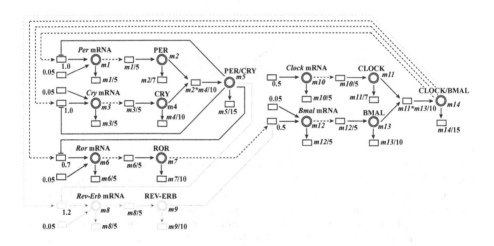

Fig. 5. HFPN model of the Fig. 4. The interactions of *Ror* and its product are added. The pale color part represents the interactions of *Rev-Erb* and its products. A simulation without this pale color part is conducted to examine a behavior of the model without *Rev-Erb* interactions.

5   Genetical Results Suggesting a Functional Interaction Between *Cry* and *Ror*

Our computer simulation analyses on the circadian clock system using HFPN presented above showed that, if cooperative interaction of CRY and ROR is

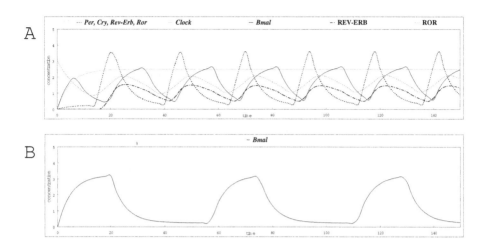

Fig. 6. (A) The behaviors of *Cry*, *Bmal*, REV-ERB, and ROR are obtained from the simulation with the HFPN shown in Fig. 5 that is without the pale part that describes *Rev-Erb* interactions. **(B)** *Bmal* behavior in the simulation with the modified HFPN where the two arcs directed toward the continuous place *Cry mRNA* are removed. This simulation result contradicts the result of biological experiment that confirmed no oscillation in *Bmal* expression in *Cry* mutant mice [24].

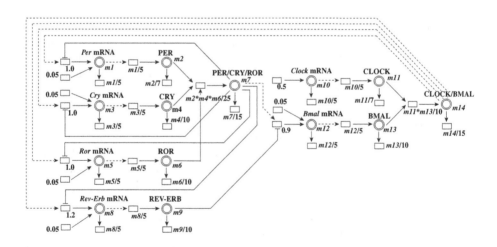

Fig. 7. HFPN model incorporating the interaction between ROR and CRY. An amount is provided to place PER/CRY/ROR from the continuous transition to which three normal arcs are directed from places PER, CRY, and ROR. The test arc from PER/CRY/ROR activates the continuous transition at its head, which works to increase the amount in place *Bmal* mRNA.

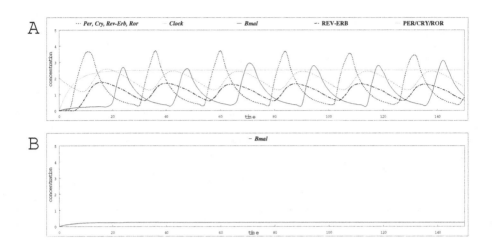

Fig. 8. Simulation results of the HFPN model in Fig. 7. **(A)** The *Bmal* peak is present approximately between (mid point) of successive *Cry* peaks. **(B)** No oscillation in *Bmal* expression in *Cry* mutant mice. Both these results correspond with the results of biological experiments.

working, most characteristics of the molecular circadian clock could be reproduced in silico. Hence, we also explored genetically for evidence suggesting *Bmal* transcription rate is upregulated both by ROR and CRY. We determined bioluminescence from *Bmal* transcription in NIH3T3 cells in culture.

NIH3T3 cells are a well-used cell line, which are known to contain the functional cell-autonomous circadian clock system. NIH3T3 cells were cultured in 24-well plates and transfected as described previously [11]. The total amount of DNA was adjusted to 400 ng per well with pcDNA3 plasmid. The cells were immediately frozen in liquid nitrogen and stored at -80°C. The cell lysates were used for the Dual Luciferase Assay System (Promega), as described previously [11].

Following the procedures above, we constructed the *Bmal1* promoter connected with luciferase gene and transfected this construct into NIH3T3 cells, in order to measure the rate of transcription of *Bmal1* gene by means of bioluminescence from luciferase. Details of the experimental methods have been published in the paper [11].

Fig. 9 shows that *Bmal1* transcription in terms of bioluminescence. We found that CRY expression induced transcriptional activation of the *Bmal1* gene (Fig. 9A), and that this activation was completely dependent on the ROR responsive elements (Fig. 9B), indicating that CRY-mediated activation of the *Bmal1* gene was via endogenous RORα. Although even RORα4 alone induces about 3 fold increase in *Bmal1* transcription as previously reported, simultaneous incubation of CRY and RORα4 increases the induction of *Bmal1* more than three times than those with CRY alone (Fig. 9C). These results clearly indicate the presence of cooperative interaction between ROR and CRY as far as *Bmal1* transcription is concerned.

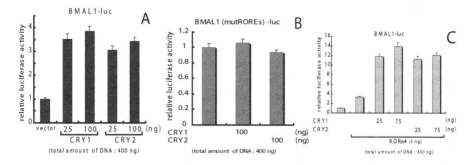

Fig. 9. (A-C) NIH3T3 cells were transfected with the indicated combinations of expression vectors. The effect of CRY on *Bmal1* transcription (A), the effect of CRY on transcriptional levels of the Bmal1 promoter carrying mutations in the ROR response elements (B), and the effect of ROR α4 and CRY on *Bmal1* transcription (C) were evaluated using a luciferase assay. Data represent the mean ± SE of triplicate samples. Numbers (under the abscissa axis) indicate the amount of DNA transfected into NIH3T3 cells. The total amount of DNA was adjusted to 400 ng with pcDNA3 plasmid.

6 Discussion

REV-ERBα has been suggested as the major regulator that drives the antiphasic transcription cycles between *Bmal* and *Cry* in mammalian circadian rhythms [3]. However, a computer simulation in a previous study [21] gave a contradicted sugession to this, showing that these antiphasic transcription cycles cannot be regulated by REV-ERBα alone. Although, in this paper, a hypothetical path "PER/CRY activates *Bmal1*" was added to realize these antiphasic cycles, this path has not been identified yet.

The *Ror* incorporated HFPN model of Fig. 4 produced antiphasic *Bmal* and *Cry* expressions. This strongly implied that the gene responsible for the antiphasic nature of *Bmal1* and *Cry* transcriptions is not *Rev-Erbα* but *Rorα*. However, this new simulation still retained an inconsistency in that *Bmal* maintains the oscillation in its expression even when *Cry* expression is disrupted. This inconsistency was eliminated in the subsequent simulation in which the interaction among PER, CRY and RORα was introduced. This simulation result, together with the fact that the REV-ERB/ROR response element is present upstream of *Cry*, suggested the existence of an undiscovered interaction between CRY and ROR that affects *Bmal* transcription. This suggestion was supported by a biological experiment that shows the synergistic activation of *Bmal* by CRY and ROR.

Post-translational modifications are important regulatory factors in controlling the circadian rhythms. In recent papers, the balance between two proteins, CK1δ/ϵ and PP1, was shown to determine the period of circadian cycle through the regulation of the kinetics of PER phoshorylation [26], and FBXL3 and FBXL21 have been shown as combinatory regulators that stabilize CRY oscillation by exerting antagonizing actions in ubiquitination on CRY [27]. These post-translational modifications need to be included in our future model that allows us to produce more precise and reliable prediction for the regulation of the mammalian circadian cycle.

Acknowledgements. The authors would like to express our gratitude to a sometime professor Shin-Ichi T. Inouye at Yamaguchi University for his constructive comments and warm encouragement on this work. Also, one of the authors HM would like to thank Mr. Yasushi Fujii who had been working on the modeling of circadian molecular reguatory system as his graduate study. This work was partially supported by Grants-in-Aid for Scientific Research (B) Number 22241050 from Japan Society for the Promotion of Science.

References

1. Dunlap, J.C.: Molecular bases for circadian clocks. Cell 96, 271–290 (1999)
2. Shearman, L.P., Sriam, S., Weaver, D.R., Maywood, E.S., Chaves, I., Zheng, B., Kume, K., Lee, C.C., van der Horst, G.T., Hastings, M.H., Reppert, S.M.: Interacting molecular loops in the mammalian circadian clock. Science 288, 1013–1019 (2000)
3. Preitner, N., Damiola, F., Lopez-Molina, L., Zakany, J., Duboule, D., Albrecht, U., Schibler, U.: The orphan nuclear receptor REV-ERBα controls circadian transcription within the positive limb of the mammalian circadian oscillator. Cell 110, 251–260 (2002)
4. Goldbeter, A.: Computational approaches to cellular rhythms. Nature 420, 238–245 (2002)
5. Leloup, J., Goldbeter, A.: Toward a detailed computational model of the mammalian circadian clock. Proc. Natl. Acad. Sci. USA 100, 7051–7056 (2003)
6. Forger, D.B., Peskin, C.S.: A detailed predictive model of the mammalian circadian clock. Proc. Natl. Acad. Sci. USA 100, 14806–14811 (2003)
7. Becker-Weimann, S., Wolf, J., Kramer, A., Herzel, H.: A model of the mammalian circadian oscillator including the REV-ERBα module. Genome Inform. 15, 3–12 (2004)
8. Leloup, J., Goldbeter, A.: Modeling the circadian clock: from molecular mechanism to physiological disorders. BioEssays 30, 590–600 (2008)
9. Becker-Weimann, S., Wolf, J., Herzel, H., Kramer, A.: Modeling feedback loops of the mammalian circadian oscillator. Biophysical J. 87, 3023–3034 (2004)
10. Sato, T.K., Panda, S., Miraglia, L.J., Reyes, T.M., Rudic, R.D., McNamara, P., Naik, K.A., FitzGerald, G.A., Kay, S.A., Hogenesch, J.B.: A functional genomics strategy reveals Rora as a component of the mammalian circadian clock. Neuron 43, 527–537 (2004)
11. Akashi, M., Takumi, T.: The orphan nuclear receptor RORα regulates circadian transcription of the mammalian core-clock Bmal1. Nat. Struc. Mol. Biol. 12, 441–448 (2005)
12. Guillaumond, F., Dardente, H., Giguere, V., Germakian, N.: Differential control of Bmal1 circadian transcription by REV-ERB and ROR nuclear receptors. J. Biol. Rhythms 20, 391–403 (2005)
13. Liu, A.C., Tran, H.G., Zhang, E.E., Priest, A.A., Welsh, D.K., Kay, S.A.: Redundant function of REV-ERVα and β non-essential role of Bmal1 cycling in transcriptional regulation of intracelluar circadian rhythms. PLoS Genetics 4(2), e10000023 (2008)

14. Mirsky, H.P., Liu, A.C., Welsh, D.K., Kay, S.A., Doyle, F.J.: A model of the cell-autonomous mammalian circadian clock. Proc. Natl. Acad. Sci. USA 106, 11107–11112 (2009)
15. Korenčič, A., Bordyugov, G., Košir, R., Rozman, D., Goličnik, M.: The interplay of *cis*-regulatory elements rules circadian rhythms in mouse liver. PLOS ONE 7(11), e46835 (2012)
16. Relógio, A., Westermark, P.O., Wallach, T., Schellenberg, K., Kramer, A.: Tuning the mammalian circadian clock: Robust synergy of two loops. PLoS Comput. Biol. 7(12), e1002309 (2011)
17. Takeda, Y., Jothi, R., Birault, V., Jetten, A.M.: RORγ directly regulates the circadian expression of clock genes and downstream targets in vivo. Nucleic Acid Research 40, 8519–8535 (2012)
18. Solt, L.A., Burris, T.P.: Action of RORs and their ligands in (patho)physiology. Trends in Endocrinology and Metabolism 23, 619–627 (2012)
19. Du, J., Xu, R.: RORα, a potental tumor supressor and therapeutic target of breast cancer. Int. J. Mol. Sci. 26, 15755–15766 (2012)
20. Matsuno, H., Tanaka, Y., Aoshima, H., Doi, A., Matsui, M., Miyano, S.: Biopathways representation and simulation on hybrid functional Petri net. In Silico Biol. 3, 389–404 (2003)
21. Matsuno, H., Inouye, S.T., Okitsu, Y., Fujii, Y., Miyano, S.: A new regulatory interaction suggested by simulations for circadian genetic control mechanism in mammals. J. Bioinform. Comput. Biol. 4, 139–153 (2006)
22. Cell Illustrator, `http://www.cellillustrator.com`
23. Reppert, S.M., Weaver, D.R.: Coordination of circadian timing in mammals. Nature 418, 935–941 (2002)
24. Oishi, K., Fukui, H., Ishida, N.: Rhythmic expression of BMAL1 mRNA is altered in Clock mutant mice: differential regulation in the suprachiasmatic nucleus and peripheral tissues. Biochem. Biophys. Res. Commun. 268, 164–171 (2000)
25. Onishi, H., Yamaguchi, S., Yagita, K., Ishida, Y., Dong, X., Kimura, H., Jing, Z., Ohara, H., Okamura, H.: Rev-erbalpha gene expression in the mouse brain with special emphasis on its circadian profiles in the suprachiasmatic nucleus. J. Neurosci. Res. 68, 551–557 (2002)
26. Lee, H., Chen, R., Kin, H., Etchegaray, J., Weaver, D.R., Lee, C.: The period of the circadian oscillator is primarily determined by the balance between casein kinase 1 an protein phosphatase 1. Proc. Natl. Acad. Sci. USA 108, 16451–16456 (2011)
27. Hirano, A., Yumimoto, K., Tsunematsu, R., Matsumoto, M., Oyama, M., Kozuka-Hata, H., Nagasawa, T., Lanjakornsiripan, D., Nakayama, K.I., Fukada, Y.: FBXL21 regulates oscillation of the circadian clock through ubiquitination and stabilization of Cryptochromes. Cell 152, 1106–1118 (2013)

HybHap: A Fast and Accurate Hybrid Approach for Haplotype Inference on Large Datasets

Rogério S. Rosa and Katia S. Guimarães

Informatics Center
Federal University of Pernambuco, UFPE
Recife, Brazil
{rsr,katiag}@cin.ufpe.br

Abstract. We introduce HybHap, a new approach for haplotype inference problem on large genotype datasets. HybHap is a hybrid method, based on the Parsimonious tree-grow idea, which resorts to Markov chains, in order to maximize the probability that the haplotypes will be shared by more genotypes in the dataset. Several experiments with large biological datasets taken from HapMap were performed to compare HybHap with two well known algorithms: fastPHASE and PTG. The results show that HybHap is a rather robust, reliable, and efficient method that runs orders of magnitude faster than the others, producing results of comparable accuracy, hence being much more suitable to deal with the challenge of genome wide tasks.

Keywords: haplotype inference, hybrid algorithms, markov chains, tree-grow.

1 Introduction

An alteration of one isolated nucleotide base which occurs with considerable frequence in the DNA of a given population is known as Single Nucleotide Polymorphism (SNP) [1] [2] [3]. Occurrences of SNPs have been associated with specific phenotypic traits and also with several illnesses [4]. Hence, it is important to map the occurrences of SNPs, but that has shown to be a huge challenge.

An haplotype can be defined as a set of SNPs from a copy of a specific chromosome. Much of the difficulty of finding these alterations is due to the lack of haplotype data in large scale, mostly because of the high cost of collecting that information directly.

One possible way of acquiring haplotype data is to infer them from genotype data, which are highly abundant. That motivates the Haplotype Inference (HI) problem, whose computational cost depends on the evolutionary model considered. One such model is based on the biological sound Parsimony Principle, but it is proved to be NP-hard [5], meaning that all algorithms currently known can only solve it in time that is exponential on the number and size of the DNA sequences, which is prohibitive. Several computational methods were developed aiming at finding solutions that may be biologically plausible, but they usually

J.C. Setubal and N.F. Almeida (Eds.): BSB 2013, LNBI 8213, pp. 24–35, 2013.
© Springer International Publishing Switzerland 2013

present high computational costs. In view of real applications, the current challenge is to infer haplotypes from large scale genotypes. Hence, computationally efficient methods with acceptable accuracy are in great demand. In this paper we present a hybrid approach that combines the efficiency of Parsimonious Tree-Grow with Markov chain choices. The result is a method that is orders of magnitude faster than the known methods, delivering results of comparable accuracy.

The rest of this paper is organized as follows. Section 2 gives a formal definition of the Haplotype Inference Problem and discusses related works. In Section 3, the proposed Markov chain used by the HybHap method is presented. The HybHap approach is introduced in Section 4. In Section 5 there is a description of the datasets and how the benchmark was organized, and in Section 6, the results of experiments are provided to demonstrate the accuracy and efficiency of the proposed hybrid method. Finally, we present several remarks, concluding the paper in Section 7.

2   Haplotype Inference Methods

We adopt the notation used by Rosa and Guimarães [6]. A genotype can be computationally represented by a vector on the alphabet $\{0, 1, 2\}$, where a symbol 2 represents an ambiguous site. Then a genotype vector g, with n sites, can be explained by two haplotype vectors h_1 and h_2, where each site $h_1(i)$ and $h_2(i)$, $1 \leq i \leq n$, has $h_1(i)$, $h_2(i) \in \{0,1\}$, and follows the rule given by: (A) $h_1(i) = h_2(i) = g(i)$, if $g(i) \in \{0,1\}$; and (B) $h_1(i) = 1 - h_2(i)$, if $g(i) = 2$. The sites of g that have a symbol 0 or 1 are called homozygous (non ambiguous sites) and those with a symbol 2 are called heterozygous (ambiguous sites). The Haplotype Inference Problem basically consists of finding, for each genotype g, haplotypes h_1 and h_2 such that h_1 and h_2 explain g in a biologically plausible way. For instance, if $g = (0, 1, 2, 2, 1, 2)$, possible solutions are $h_1 = (0, 1, 0, 0, 1, 0)$ and $h_2 = (0, 1, 1, 1, 1, 1)$, or else $h_1 = (0, 1, 0, 1, 1, 1)$ and $h_2 = (0, 1, 1, 0, 1, 0)$, among other possibilities. It is easy to see that there are 2^{h-1} candidate haplotype pairs to explain g, where h is the number of ambiguous sites in g. Obviously, there are many plausible solutions for a given input g, so a biological criterion is needed to define a good solution.

There are two main biological models used to infer haplotypes: Pure Parsimony and Perfect Phylogeny. Inferring haplotypes assuming perfect phylogeny was shown to be a linear problem [7]. However the assumption that the DNA sequences were not subject to recombination events is not realistic.

Haplotype inference by pure parsimony principle (HIPP) has been used by many approaches because of its innate simplicity and biological soundness. As said before, unfortunately the HIPP problem is NP-hard [8]. Some approaches based on Integer Programming have been proposed for it [9] [10] [11] [12].

Another method for the HIPP problem is the Parsimonious Tree-Grow (PTG) method [13], which explains a set of m genotypes of length n in time $O(m^2 n)$. In the PTG method a tree is constructed, where each edge is labelled by a haplotype symbol (0 or 1), and nodes contain the genotypes (id) that are explained by haplotypes formatted by a trace from the root to that specific node. Many operations in PTG are random, so it is necessary to run the method many times, selecting the best solution using some metric, in order to have reliable results.

Methods based on Markov chain models have been proposed successfully. These methods basically build a Markov chain in which each state is associated to a symbol (0 or 1) and the transition probabilities are calculated from the input data. Heuristics based on Dynamic Programming and Expectation Maximization algorithms are also applied [14] [15] [16] [17].

Statistical methods have considered the Parsimony Principle as accessory. PHASE [18] and fastPHASE [19] are considered good classical approaches for the HI Problem. These methods use maximum likelihood to estimate haplotype frequencies. The objective is to estimate the maximum value of this likelihood function. Such methods are stochastic, and each execution of the program may result in a different solution, since the derivations are dependent on the initial configuration, which is randomly selected. Basically, fastPHASE is a variation of PHASE for resolving large data sets.

3   Computing the Markov Chain

The probability that a haplotype fragment will be part of a solution, considering the parsimony criterion can be efficiently estimated using a Markov chain. Given a genotype matrix G, with m rows and n columns, a Markov chain C is created with $2n+2$ states, each state representing the start (C_{start}) or the end of the chain (C_{end}), or a possible symbol s (0 or 1) in the j-th site of a haplotype fragment, $1 \leq j \leq n$, ($C_j(s)$). There are three types of state transitions: ($C_{start}, C_1(s)$), ($C_{j-1}(s_1), C_j(s_2)$), and ($C_n(s), C_{end}$).

The initial probabilities are computed as an *a priori* probability of symbol s occurring in the first site of all the $2m$ haplotypes to be inferred from G (1). The absolute frequency of symbol s being in the first site of the matrix is calculated according to Equation 2.

$$(C_{start}, C_1(s)) = A(1, s)/(2m) \tag{1}$$

$$A(j, s) = \sum_{y=1}^{m} f_1(G(y, j), s), \tag{2}$$

with

$$f_1(x, s) = \begin{cases} 2, & \text{if } x = s \\ 1, & \text{if } x = 2 \\ 0, & \text{otherwise} \end{cases}$$

The transition probabilities whose source state is not the initial state (C_{start}) and the destination is not the final state (C_{end}) are denoted by ($C_{j-1}(s_1), C_j(s_2)$), where $2 \leq j \leq n$. These are conditional probabilities: probability of s_2 occurring in the j-th site of the $2m$ haplotypes inferred from G, given that s_1 occurred in the $(j - 1)$-th site of said set of haplotypes (Equation 3). That depends on the absolute frequency of haplotypes inferred with s_1 in the $(j - 1)$-th site (1), and an estimation of the expected frequency of symbol s_2 in the j-th site of those same haplotypes (Equation 4).

$$(C_{j-1}(s_1), C_j(s_2)) = B(j, s_1, s_2)/A(j - 1, s_1) \tag{3}$$

$$B(j, s_1, s_2) = \sum_{y=1}^{m} f_2(G(y, j - 1), G(y, j), s_1, s_2), \tag{4}$$

with

$$f_2(x_1, x_2, s_1, s_2) = \begin{cases} 2, & \text{if } x_1 = s_1 \text{ and } x_2 = s_2 \\ 0.5, & \text{if } x_1 = 2 \text{ and } x_2 = 2 \\ 1, & \text{if } x_1 = 2 \text{ and } x_2 = s_2 \\ 1, & \text{if } x_1 = s_1 \text{ and } x_2 = 2 \\ 0, & \text{otherwise} \end{cases}$$

After constructing Markov chain C as described above, a tree T is computed which contains the $2m$ haplotypes that resolve G. The HybHap method uses the information contained in C to choose more promising branches, trying to keep T with the minimum possible number of branches, so as to approach an optimal solution, according to the Pure Parsimony criterion.

4   The HybHap Method

A tree T, which has $n+1$ layers, each one denoted by $T(j)$, is computed. A layer can have $2m$ nodes in the worst case (maximum possible number of distinct haplotypes to be inferred). A node in layer $T(j)$ is denoted by $T_{(j,k)(s)}$, where s is the node type (0 or 1), and k is the number sequence of the node in layer j.

A node $T_{(j,k)(s)}$ is labelled by $(i_{r_1}, i_{r_2}, ..., i_{r_g})$, $1 \leq g \leq m$, which represents the genotype fragments explained by that node. The root is labelled by all genotype Ids $(i_1, i_2, ..., i_m)$. For each layer j of T, for each node in j, for each genotype Id i_r in the label of the current node, if G, in site j of genotype, has value 1 (0), then a node of type 1 (0) is created on the next layer connected to the current node, and it is labelled by all genotype Ids in the present node that have value 1 (0) in site j. If the value in G in layer j and genotype Id i_r has value 2, then it is checked if a value 2 was previously explained for genotype i_r. If that is not the case, then site j in genotype i_r is resolved by adding two new nodes in the $(j+1)$-th layer, connected to the current node, and labelled i_r. In case a value 2 has been previously resolved, then the genotype Id and the current node are reserved to be processed after the current layer is treated. An example of tree is illustrated in Figure 1.

Random operations may occur in the processing of the genotypes and nodes reserved in a layer. In HybHap the Markov chain C constructed will be used to decide which haplotype fragment is the most promising one. Each trace, from

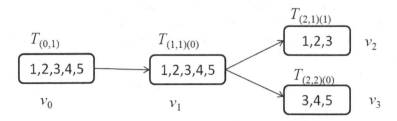

Fig. 1. Node v_0 is the root of tree T; this node is in level 0 and is node number 1 of that level ($T_{(0,1)}$). The root is labelled by five genotypes Ids (1,2,3,4, and 5), representing all five genotypes in the input. Node v_1, denoted by $T_{(1,1)(0)}$ is in level 1 and it is the node number 1 of that level; this node is of type 0. Node v_3, denoted by $T_{(2,2)(0)}$, is in level 2 and it is the node number 2 of that level, this node is of type 0, and it is labelled by genotype Ids 3,4, and 5.

the root to the current node in T, is a valid path in C. There are three situations in which random choices may be needed, the others are symmetric; in those situations, new nodes are computed through C, and the choice is based on the maximum probability found. In case we need to choose among existing nodes to explain the reserved genotype, then we compute the Euclidean distance between the sites that have not yet been processed in the reserved genotype and all sites that have not been processed in the genotypes that are partially resolved by candidate nodes, the choice is based on the least distance. When the probabilities or distances are the same between candidate nodes, then a random choice is needed, but the chances of that actually occurring are slim.

The HybHap method (Algorithm 1) has three main steps: Initialization, Resolution of genotype prefix with known solution (genotype fragments that have only homozygous sites or one heterozygous site), as described in Algorithm 2, and explanation of genotype fragments that have no previous resolution (more than one heterozygous site), as described in Algorithm 3. In initialization the Markov chain is computed as described before, the root is created and labelled with all genotype Ids of G.

The 2 explains the genotype fragments (prefix) that have at most one heterozygous site. In this case, when a site with symbol 2 is resolved for genotype i, we make $f(i) = true$. All genotypes marked in the prior step ($f(i) = true$), for a specific SNP (a layer of tree T), will be processed after all non-ambiguous genotype fragments of that layer are resolved.

In Algorithm 3, the fragments of genotypes reserved before are explained. In 2, a genotype i was associated to two nodes (two is the maximum number of nodes that can explain a genotype with at least one heterozygous site). In this Algorithm 3, a Markov chain is used to decide which is the best branching option. Equation 5 is used, in which $P(v_1)$ denotes the probability that the haplotype fragment represented by node v_1 will be part of solution, according to the parsimony criterion (conservation), and $t(v_1)$ denotes the type of v_1. There are three cases in which the Markov chain is applied: (1) There are no branches growing from v_1 and v_2; (2) There is a single branch (v'_1, v'_2) growing from each of v_1 and v_2, both of the same type s; and (3) There are two branches growing from v_1 but no branches growing from v_2; the other cases are symmetric. Those three situations are addressed in Algorithm 3, and illustrated in Figure 2-C.

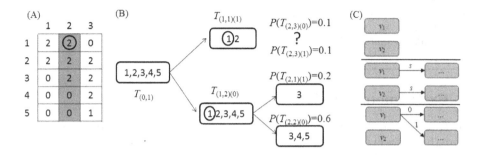

Fig. 2. Example of Algorithm Execution

$$P(v_1)(C_j(t(v_1)), C_{j+1}(s_1)) + P(v_2)(C_j(t(v_2)), C_{j+1}(s_2)) \qquad (5)$$

After building tree T, the final solution can be recovered by tracing from the root to each leaf of T, concatenating the types of the nodes in the path. The result will be the haplotype matrix that explains G following the parsimony principle.

Figure 2-B illustrates an application of Markov chain C during the construction of tree T. First the root $T_{(0,1)}$ is created and labelled by all genotype Ids of the matrix identified in Figure 2-A. Then the nodes descendent from $T_{(0,1)}$ are created. Since no ambiguity has been resolved in T yet, nodes $T_{(1,1)(1)}$ and $T_{(1,2)(0)}$ are created to explain genotypes (1,2) and (1,2,3,4,5), respectively. Since in column 2, genotypes 1 and 2 have symbol 2, and sites of that type have been previously resolved for those genotype, their Ids are kept to be processed after all sites on the second column that do not present ambiguity or that have all previous sites without ambiguity are resolved. Hence, genotypes 3, 4, and 5 are resolved, by creating nodes $T_{(2,1)(1)}$ and $T_{(2,2)(0)}$.

After that, genotypes 1 and 2 are dealt with. There are two nodes, $T_{(2,1)(1)}$ and $T_{(2,2)(0)}$, branching from $T_{(1,2)(0)}$ that can explain genotype 1, and none from $T_{(1,1)(1)}$. In order to decide which of those nodes should be created branching from $T_{(1,1)(1)}$: $T_{(2,3)(0)}$ or $T_{(2,3)(1)}$. Markov chain C is then used to estimate which node maximizes the probability of being also used in the resolution of other genotypes (parsimony). In the case of this example, there are the following node combinations: $P(T_{(2,3)(0)} + T_{(2,1)(1)}) = 0.3$ and $P(T_{(2,3)(1)} + T_{(2,2)(0)}) = 0.7$. The choice is for the option that maximizes the probability, hence, node $T_{(2,3)(1)}$ is added branching from node $T_{(1,1)(1)}$.

Algorithm 1. HybHap

---

    input : a matrix of Genotypes G
    output: a tree T
1 Initialization;
2 **foreach** *layer j in T* **do**
3   **foreach** *node v in current layer* **do**
4 | KnownSolution(v, j);
5   **end**
6   UnknownSolution(j);
7 **end**

---

5   Experiments Design

The dataset used for the experiments was the same one used in a previous work where the performances of well known haplotype inference algorithms when dealing with data of different sizes and levels of conservation are compared [6]. It is comprised by sequences originally taken from the HapMap Project [20], which were collected from Chromosome 20 of population CEU (Caucasians resident in the state of Utah (USA) with northern European ancestry). The original dataset is composed of 13 subsets which vary in sequence length and in number of distinct haplotypes.

The set contains haplotypes of sizes 100, 200, 400, 800, 1600, and 3200 SNPs with 88 individuals, separated by size into classes A, B, C, D, E, and F, respectively. From each class, we chose randomly three instances.

The metrics used in the benchmark were Error Rate [21] and computational time. The Error Rate tells us about the capacity that one method has to correctly infer a haplotype set from a genotype set, based on a known haplotype set. The computational time is an empiric metric used to estimate computational costs; although it is not the best technique for it, in this case theoretical analysis cannot be applied to all methods.

For the comparison experiments, PTG was implemented in MATLAB 2008. Version 1.2.3 for Windows of fastPHASE was used. The experiments ran individually in a computer with an Intel Quad Core 2.33GHz processor, with 3GB of RAM. The results are shown in Table 1. For each experiment, the execution time (Time) and Error Rate (ER) attained are given.

6   Experiments Results

The measures described earlier were applied to each instance. Since PTG and fast-PHASE have a stochastic behavior, for comparison purposes the average over 30 executions with every single dataset was used to establish the Error Rate. Although HybHap is not a deterministic algorithm, in practice it presents standard deviation virtually equal to zero, meaning that for different executions with the same input dataset (including the same genotype order), it generates the same haplotype set.

Comparing HybHap to PTG, considering Error Rate, the accuracy of HybHap and PTG were very close, slightly favoring HybHap for the larger datasets. In all

---

Algorithm 2. KnownSolution(v: a node, j: a SNP)

---

  1 **foreach** *genotype i in v* **do**
  2 | **if** *$G(i,j+1)=2$ and $f(i)=true$* **then**
  3 | | Associate to genotype i the node v;
  4 | **else**
  5 | | **if** *$G(i, j + 1) = s$ and $s \in \{0,1\}$* **then**
  6 | | | **if** *there is no branch with target node of type s growing from node v*
           | | | **then**
  7 | | | | Add a node of type s growing from node v and label this new
           | | | | node with i;
  8 | | | **else**
  9 | | | | Add i to the set of Ids in the label of this node;
 10 | | | **end**
 11 | | **else**
 12 | | | $f(i) \leftarrow true$;
 13 | | | **for** $s = 0, 1$ **do**
 14 | | | | **if** *there is no node growing from v of type s* **then**
 15 | | | | | Add a new node from v in layer $j + 1$ of type s and add i to
           | | | | | the set of Ids of this new node;
 16 | | | | **else**
 17 | | | | | Add i to the set of Ids of this node;
 18 | | | | **end**
 19 | | | **end**
 20 | | **end**
 21 | **end**
 22 **end**

---

cases, HybHap was much faster than PTG. For instance, HybHap solved the largest dataset (F) in about 3 minutes, while PTG took about 39 minutes to find a less accurate solution.

Comparing HybHap to the classical approach fastPHASE, we observed that the accuracy performances considering Error Rate were very close, and for the larger datasets in the benchmark, the differences between the Error Rates for the two methods were smaller than 2%. It is important to notice that, for the largest dataset, F, while HybHap needed only about 1 minute to find a solution with 13.67% of error, fastPHASE resolved this instance with 11.93% of error in about 72 hours. The difference of Error Rate in this case was 1.74%, however, the time necessary for fastPHASE to resolve it was approximately 1080 times longer than the time required by HybHap.

Figure 3 shows graphical comparisons of HybHap with fastPHASE and PTG, in regard to Error Rate (Figure 3-A) and computational time (Figure 3-B). Since the values for computational time are so different, the values in Figure 3-B are depicted in log scale. It can be seen that the time of fastPHASE grows much faster than HybHap, as the length of the sequences in the datasets increases. On the other hand, while the Error Rate of HybHap is always higher than that of fastPHASE, the difference in Error Rate is virtually constant.

Algorithm 3. UnknownSolution(j: a SNP)

---

1 **foreach** *genotype i associated to a node pair (v_1, v_2) in SNP j* **do**
2   **if** *there is a single branch growing from v_1 and v_2 of different types or there are two branches and a single branch growing from v_1 or v_2* **then**
3   Add i to the set of Ids of a node that grows from a node that has a single branch (v_1 or v_2) and add i to the set of Ids of a node of opposite type that grows from the node left;
4   **else**
5   **if** *there are two branches growing from v_1 and two branches growing from v_2* **then**
6   Compute the Euclidean distance among the unresolved suffix of i and the unresolved suffixes of genotypes explained by nodes growing from v_1 and v_2. Explain i in nodes that minimize the distance.
7   **else**
8   Compute value of Equation 5 for $(s_1 = 0, s_2 = 1)$ and for $(s_1 = 1, s_2 = 0)$. Take the pair (s_1, s_2) that maximizes v, if value of the two pairs are same, then select a pair randomly;

9   **switch** *Ambiguity cases in v_1 and v_2* **do**
10   **case** *1*
11   Grow a node of type s_1 from $v1$, Grow a node of type s_2 from v_2. Add i to the set of Ids of these new nodes;

12
13   **case** *2*
14   If $s = s_1$, then add genotype i to the set of node v_1' and grow a node of type s_2 from v_2, labelled by i. Otherwise, do the same symmetrically;

15
16   **case** *3*
17   Add genotype i to the set of Ids of the node of type s_1 growing from v_1, and add a node of type s_2 growing from node v_2, including i in the set of Ids of this new node (other cases are symmetric);

18
19   **endsw**
20   **end**
21   **end**
22 **end**

Table 1. Comparison Results: Error Rate (ER) and Time in seconds (s), minutes (m) or hours (h)

Set	HybHap		PTG		fastPHASE	
	ER	Time	ER	Time	ER	Time
A	12.9%	02.88 s	13.5%	53.88 s	09.3%	00 h 30 m
B	09.9%	04.83 s	09.7%	01.62 m	05.8%	01 h 00 m
C	12.7%	12.28 s	12.6%	03.73 m	09.2%	02 h 30 m
D	11.1%	26.11 s	12.3%	07.67 m	08.7%	05 h 24 m
E	13.6%	01.06 m	14.2%	16.98 m	11.7%	18 h 00 m
F	13.7%	03.00 m	14.2%	39.00 m	11.9%	72 h 00 m

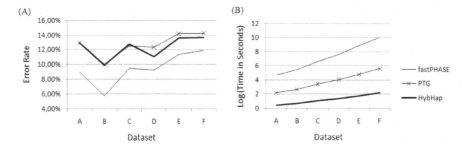

Fig. 3. Computational Time (in seconds) and Error Rate attained in each dataset class. Each class has 88 genotypes and different number of SNPs: (A) 100, (B) 200, (C) 400, (D) 800 and (F) 1600.

7 Discussion and Conclusion

In this paper we have proposed a hybrid method for haplotype inference. The proposed method is very stable, since in practice it presents a standard deviation of zero. In our experiments, the highest number of random operations for any instance was two, but that seldom happened. Due to that and to the efficiency of the operations in HybHap, its computational time is very low when compared to PTG and to fastPHASE.

With the enormous growth in the number of genomes available, efficient methods to deal with large datasets are highly desirable. There are many approaches to infer haplotypes with high quality, but they are applied only to small datasets, and it is not in line with the current inference requirements, which are on large scale. In face of that, HybHap presents desirable properties. The proposed method is computationally very efficient and in large datasets produces results with accuracy very close to that of more costly methods. That is most valuable, due to the growing number of genetic variation studies, which are performed by Computational Biology groups most of which have limited computational processing resources available.

An important point is the fact that PTG is based solely on parsimony, disregarding any other type of information or precondition about the genotype sequences, while methods based on Markov chains, such as fastPHASE, use the parsimony criterion as a help, applying additional techniques, models, and insights to find a biologically more plausible solution. Nonetheless, that combination leads to an extremely high computational time requirement. Hence, as the length of the genotype sequences grow, those methods become non-viable.

We also believe that the wide gap between the performances of HybHap and fastPHASE with respect to some cases of Error Rate is due to the fact that HybHap has no strategy to cluster together segments of different with similar characteristics regarding conservation. Since HybHap presents excellent computational cost, the original algorithm, presented in this paper, can be improved by strategies to associate similar regions, as it is done in fastPHASE, for instance. We are currently working on that aspect of the method. Missing data is another aspect that needs to be addressed.

The experimental analysis shows that the HybHap method is more adequate for dealing with long genome sequences. We are currently working on a theoretical argument for the fact that HybHap requires less computational time, as well as on improving its accuracy.

Acknowledgments. This work was developed with financial support from Brazilian sponsoring agency FACEPE (process no. PBPG-0070-1.03/10). The authors thank the anonymous reviewers for their insightful comments.

References

1. Sonis, S., Antin, J., Tedaldi, M., Alterovitz, G.: SNP-based Bayesian networks can predict oral mucositis risk in autologous stem cell transplant recipients. Oral Dis (2013)
2. Jin, Y., Lee, C.G.L.: Single Nucleotide Polymorphisms Associated with MicroRNA Regulation. Biomolecules 3(2), 287–302 (2013)
3. Martinez-Herrero, S., Martinez, A.: Cancer protection elicited by a single nucleotide polymorphism close to the adrenomedullin gene. J. Clin. Endocrinol. Metab (2013)
4. Wang, E.Y., Liang, W.B., Zhang, L.: Association between single-nucleotide polymorphisms in interleukin-12a and risk of chronic obstructive pulmonary disease. DNA Cell Biol. 31(9), 1475–1479 (2012)
5. Gusfield, D.: Haplotyping as perfect phylogeny: Conceptual framework and efficient solutions. In: International Conference on Research in Computational Molecular Biology (RECOMB), pp. 166–175 (2002)
6. Rosa, R.S., Guimarães, K.S.: Insights on haplotype inference on large genotype datasets. In: Ferreira, C.E., Miyano, S., Stadler, P.F. (eds.) BSB 2010. LNCS (LNBI), vol. 6268, pp. 47–58. Springer, Heidelberg (2010)
7. Ding, Z., Filkov, V., Gusfield, D.: A linear-time algorithm for the perfect phylogeny haplotyping (PPH) problem. In: Miyano, S., Mesirov, J., Kasif, S., Istrail, S., Pevzner, P.A., Waterman, M. (eds.) RECOMB 2005. LNCS (LNBI), vol. 3500, pp. 585–600. Springer, Heidelberg (2005)
8. Gusfield, D.: Inference of haplotypes from samples of diploids populations: Complexity and algorithms. Journal of Computational Biology 8, 305–324 (2001)

9. Gusfield, D.: Haplotype inference by pure parsimony. In: Baeza-Yates, R., Chávez, E., Crochemore, M. (eds.) CPM 2003. LNCS, vol. 2676, pp. 144–155. Springer, Heidelberg (2003)

10. Lancia, G., Pinotti, M.C., Rizzi, R.: Haplotyping populations by pure parsimony: Complex of exact and approximation algorithms. INFORMS J. Computing 16, 348–359 (2004)

11. Halldórsson, B.V., Bafna, V., Edwards, N., Lippert, R., Yooseph, S., Istrail, S.: A survey of computational methods for determining haplotypes. In: Istrail, S., Waterman, M.S., Clark, A. (eds.) DIMACS/RECOMB Satellite Workshop 2002. LNCS (LNBI), vol. 2983, pp. 26–47. Springer, Heidelberg (2004)

12. Brown, D.G., Harrower, I.M.: Integer programming approaches to haplotype inference by pure parsimony. IEEE/ACM Trans. Comput. Biol. Bioinformatics, pp. 141–154 (2006)

13. Li, Z., Zhou, W., Zhang, X., Chen, L.: A parsimonious tree-grow method for haplotype inference. Oxford Bioinformatics 21, 3475–3481 (2005)

14. Sun, S., Greenwood, C.M., Neal, R.M.: Haplotype inference using a bayesian hidden markov model. Genetic Epidemiology 31, 937–948 (2007)

15. Zhang, J.H., Wu, L.Y., Chen, J., Zhang, X.S.: A fast haplotype inference method for large population genotype data. Computational Statistics and Data Analysis 52, 4891–4902 (2008)

16. Eronen, L., Geerts, F., Toivonen, H.: Haplorec: efficient and accurate large-scale reconstruction of haplotypes. BMC Bioinformatics 7, 542 (2006)

17. Eronen, L., Geerts, F., Toivonen, H.: A markov chain approach to reconstruction of long haplotypes. In: Pacific Symposium on Biocomputing, pp. 104–115 (2004)

18. Stephens, M., Smith, N.J., Donnelly, P.: A new statistical method for haplotype reconstruction from population data. American Journal of Human Genetics 68, 59–62 (2001)

19. Scheet, P., Stephens, M.: A fast and flexible statistical model for large-scale population genotype data: applications to inferring missing genotypes and haplotypic phase. American Journal of Human Genetics 78, 629–644 (2006)

20. The International HapMap Consortium: The international hapmap project. Nature 426, 789–796 (2003)

21. Niu, T., Qin, Z.S., Xu, X., Liu, J.S.: Bayesian haplotype inference for multiple linked single-nucleotide polymorphism. American Journal of Human Genetics 70, 157–169 (2002)

Restricted DCJ-Indel Model Revisited

Marília D.V. Braga[1] and Jens Stoye[2]

[1] Inmetro – Instituto Nacional de Metrologia, Qualidade e Tecnologia, Brazil
[2] Technische Fakultät and CeBiTec, Universität Bielefeld, Germany
mdbraga@inmetro.gov.br, jens.stoye@uni-bielefeld.de

Abstract. The Double Cut and Join (DCJ) is a generic operation representing many rearrangements that can change the organization of a genome, but not its content. For comparing two genomes with unequal contents, in addition to DCJ operations, we have to allow insertions and deletions of DNA segments. The distance in the so-called general DCJ-indel model can be exactly computed, but allows circular chromosomes to be created at intermediate steps, even if the compared genomes are linear. In this case it is more plausible to consider the restricted DCJ-indel model, in which the reincorporation of a circular chromosome has to be done immediately after its creation. This model was studied recently by da Silva *et al.* (BMC Bioinformatics 13, Suppl. 19, S14), but only an upper bound for the restricted DCJ-indel distance was provided. Here we solve an open problem posed in that paper and present a very simple proof showing that the distance, that can be computed in linear time, is always the same for both the general and the restricted DCJ-indel models. We also present a simpler algorithm for computing an optimal restricted DCJ-indel sorting scenario in $O(n \log n)$ time.

1 Introduction

Genomes can be composed of one or more chromosomes, that can be linear or circular. A good estimate of evolutionary *distance* based on whole genome comparison can be obtained by asking for the minimum number of rearrangements that are necessary to transform one genome into another one. In the literature this transformation has also been referred to as *sorting* one genome *into* another genome. A sequence of rearrangements sorting a genome A into a genome B is called *scenario*, that is *optimal* when its length is minimum. Typical rearrangements that change the organization of genomes are inversions of chromosomal segments, translocations of the ends of two linear chromosomes, and chromosome fusions and fissions.

A polynomial algorithm was proposed by Hannenhalli and Pevzner in 1995 to compute the genomic distance between two genomes with equal contents considering all mentioned rearrangements. The paper [1], however, relies on the analysis of many particular cases and is full of technical details, making it susceptible to errors [2–6]. Later the same set of rearrangements were unified in the simple Double Cut and Join (DCJ) model [7], which has become very popular over the last few years due to its general applicability and mathematical elegance [8–12].

J.C. Setubal and N.F. Almeida (Eds.): BSB 2013, LNBI 8213, pp. 36–46, 2013.

Computing the DCJ distance and finding one optimal DCJ sorting scenario can be done in linear time [8]. However, while sorting a genome into another by DCJ, circular chromosomes can appear in the intermediate steps [7, 8]. In the *general model* many circular chromosomes can coexist in some intermediate step, even if the compared genomes are composed of linear chromosomes only, such as eukaryotic nuclear genomes. To account for this fact, a restricted version of the DCJ model has been considered, where in the start and end genomes all chromosomes are linear, and whenever in an intermediate step a circular chromosome is created, it has to be reincorporated into a linear chromosome in the next step. These two consecutive DCJ operations, which create and reincorporate a circular chromosome, mimic a transposition or a block interchange [7, 13].

In Figure 1 we give examples of a general and a restricted DCJ sorting scenarios. While the general and the restricted DCJ distance are equal and can be computed in linear time [7, 8], the currently best known algorithm to find an optimal restricted sorting scenario runs in $O(n \log n)$ time [13], where n is the number of common DNA segments between the compared genomes.

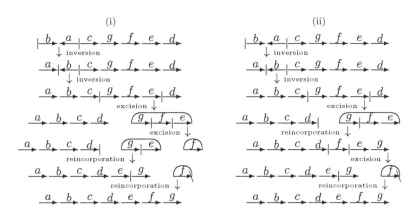

Fig. 1. (i) An optimal sorting sequence in the general DCJ model – many circular chromosomes can coexist in the intermediate species. (ii) An optimal sorting sequence in the restricted DCJ model – a circular chromosome is immediately reincorporated after its excision. The first excision-reincorporation mimics the interchange of segments d and g, while the second excision-reincorporation mimics the transposition of segment f.

Many variants of the general DCJ model have been proposed, including an extension to genomes with unequal contents, i.e. where DNA segments may be present in one, but not in the other genome. In order to transform one such genome into the other one, insertions and deletions (indels) of segments are necessary, giving rise to the so-called *DCJ-indel* model [14, 15], for which both the distance and one optimal sorting scenario can be obtained in linear time. However, like in the basic DCJ model, several circular chromosomes may coexist in intermediate steps.

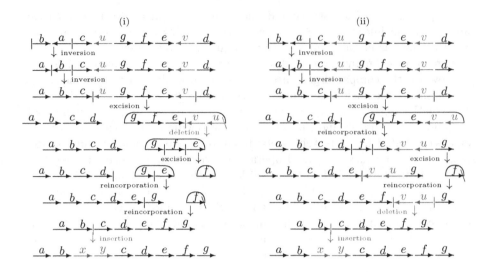

Fig. 2. (i) An optimal sorting sequence in the general DCJ-indel model – many circular chromosomes can coexist in the intermediate species. (ii) An optimal sorting sequence in the restricted DCJ-indel model – a circular chromosome is immediately reincorporated after its excision. The first excision-reincorporation mimics the interchange of segments d and $\bar{u}g$, while the second excision-reincorporation mimics the transposition of segment f.

A restricted version of the DCJ-indel model [16] has also been considered, but the question whether both the general and the restricted DCJ-indel distances were the same was not so easy to answer as it was for the general DCJ model. In fact, the paper by da Silva *et al.* [16] gives only an upper bound for the restricted DCJ-indel distance and an algorithm that achieves this bound. Deriving an exact distance formula and an optimal sorting algorithm were left as open problems. In Figure 2 we give examples of a general and a restricted DCJ-indel sorting scenarios.

In this work we prove that the distance is always the same for both the general and the restricted DCJ-indel models, as already conjectured in [17]. We also give a simple algorithm for computing an optimal sorting scenario under the restricted DCJ-indel model.

This paper is organized as follows. In Section 2 we give definitions and previous results used in this work. In Section 3 we show how to compute the distance and one optimal sorting scenario in the restricted DCJ-indel model. Section 4 concludes by relating this work to other genomic distance measures and pointing out open problems concerning their restricted versions.

2 Preliminaries

Each marker g in a genome is an oriented DNA fragment, represented by the symbol g, if it is read in direct orientation, or by the symbol $\bar{g}$, if it is read in reverse orientation. Each one of the two extremities of a linear chromosome is called a *telomere*, represented by the symbol ∘. Each chromosome in a genome can then be represented by a string that can be circular, if the chromosome is circular, or linear and flanked by the symbols ∘, if the chromosome is linear.

We deal with models in which duplicated markers are not allowed. Given two genomes A and B, possibly with unequal content, let $\mathcal{G}$, $\mathcal{A}$ and $\mathcal{B}$ be three disjoint sets, such that $\mathcal{G}$ is the set of *common markers* that occur once in A and once in B, $\mathcal{A}$ is the set of markers that occur only in A and $\mathcal{B}$ is the set of markers that occur only in B. The markers in sets $\mathcal{A}$ and $\mathcal{B}$ are also called *unique markers*.

As an example, consider the linear genomes $A = \{\circ b\bar{a}c\bar{u}gfe\bar{v}d\circ\}$ and $B = \{\circ abxycdefg\circ\}$, that are the top and the bottom genomes represented in both parts of Figure 2. Here we have $\mathcal{G} = \{a, b, c, d, e, f, g\}$, $\mathcal{A} = \{u, v\}$ and $\mathcal{B} = \{x, y\}$.

2.1 DCJ Operations

A *cut* performed on a genome A separates two adjacent markers of A. A *double-cut and join* or *DCJ* applied on a genome A is the operation that performs cuts in two different positions in A, creating four open ends, and joins these open ends in a different way. As an example consider the first DCJ applied to genome $A = \{\circ b\bar{a}c\bar{u}gfe\bar{v}d\circ\}$ in Figure 2. This operation cuts before and after $b\bar{a}$, creating the segments ∘•, •$b\bar{a}$• and •$c\bar{u}gfe\bar{v}d$∘, where the symbol • represents the open ends. If we then join the first with the third and the second with the fourth open end, we obtain $A' = \{\circ a\bar{b}c\bar{u}gfe\bar{v}d\circ\}$. This DCJ corresponds to the inversion of contiguous markers $b\bar{a}$. Indeed, a DCJ operation can correspond to several rearrangements, such as an inversion, a translocation, a fusion or a fission, and also to circular excisions and reincorporations [7].

Some additional rearrangements correspond to more than one DCJ operation. A *block interchange* occurs when two segments exchange their positions. A particular case is a *transposition*, in which one of the two segments is empty. When a block interchange or a transposition affects one single chromosome it is said to be *internal*, otherwise *external*. These rearrangements require at least three distinct cuts and cannot be represented by a single DCJ operation. Instead, they can be obtained by a composition of two DCJ operations. While external block interchanges and transpositions can always be mimicked by two consecutive translocations, internal ones can only be mimicked by two DCJs if the first is a circular excision and the second is a circular reincorporation. We call such a pair of operations an *ER composition* (see Figure 3).

2.2 The General DCJ Model

In the general DCJ model the genomes have the same content and can be unichromosomal or multichromosomal, linear or circular. Given two genomes A and B

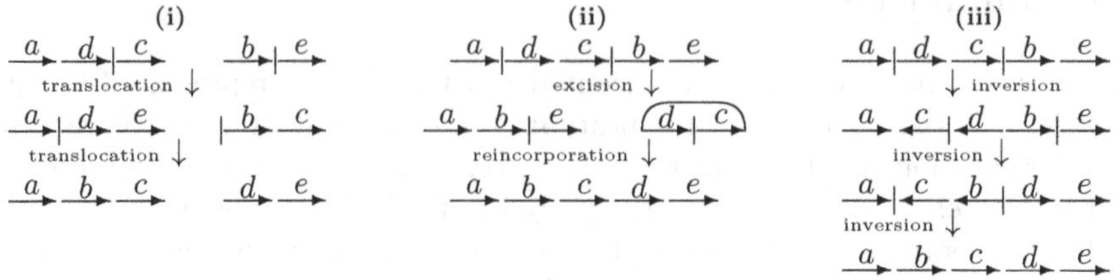

Fig. 3. (i) External block interchange of markers d and b mimicked by two translocations. (ii) Internal block interchange of markers d and b mimicked by an ER composition. (iii) Without a circular excision, the internal block interchange of markers d and b requires at least three inversions to be mimicked.

with equal contents, the *DCJ distance* of A and B, denoted by $d^{DCJ}(A,B)$, is the minimum number of DCJ operations that sort A into B and can be exactly computed in linear time [8]. Consider a DCJ ρ transforming the genome A into another genome A'. If $d^{DCJ}(A,B) = d^{DCJ}(A',B)+1$, the operation ρ is said to be optimal. Under the general DCJ model, an optimal sorting scenario, composed of optimal DCJ operations, can also be obtained in linear time [8].

The restricted DCJ model. In the restricted DCJ model the genomes are linear, unichromosomal or multichromosomal. Given two linear genomes A and B with equal contents, the *restricted DCJ distance* of A and B, denoted by $d^{rDCJ}(A,B)$, is the minimum number of DCJ operations that sort A into B, with the restriction that a circular excision has to be immediately followed by a circular reincorporation, forming an ER composition. After an optimal circular excision, there is always an optimal circular reincorporation [7]. Such an ER composition is said to be optimal and guarantees that $d^{rDCJ}(A,B) = d^{DCJ}(A,B)$. The best algorithm to find a restricted DCJ sorting scenario runs in $O(n \log n)$ time [13], where n is the number of markers in A and B, respectively.

2.3   Indels

DCJ operations are able to change only the organization of the genomes, but not their contents. When the genomes have unequal contents, we need to consider *insertions* and *deletions* of blocks of contiguous markers [14, 18]. We refer to insertions and deletions as *indel* operations. Indels have two restrictions: (i) markers of $\mathcal{G}$ cannot be deleted; and (ii) an insertion cannot produce duplicated markers [15]. At most one chromosome can be entirely deleted or inserted at once. We illustrate an indel with the following example: the insertion of markers xy into the genome $B' = \{\circ abcdefg\circ\}$, that results into $B = \{\circ abxycdefg\circ\}$, as we can see in the last sorting step of both scenarios shown in Figure 2. The opposite operation would be a deletion.

The triangular inequality problem. Since indels can be applied to blocks of markers of arbitrary size, the triangular inequality does not hold for genomic distances that consider this type of operation. Given any three genomes A, B and C and a distance measure d, consider without loss of generality that $d(A, B) \geq d(A, C)$ and $d(A, B) \geq d(B, C)$. Then the triangular inequality is the property that guarantees that $d(A, B) \leq d(A, C) + d(B, C)$.

Although this property holds for the classical models that consider only rearrangements, it does not hold for the approaches that allow indels. Consider for example the genomes $A = \{\circ abcde \circ\}$, $B = \{\circ ac\bar{d}be\circ\}$ and $C = \{\circ ae \circ\}$ [14]. While A and B can be sorted into C with only one indel, the minimum number of inversions required to sort A into B is three. In this case the triangular inequality is disrupted. This is a problem if one intends to use this distance to compute the *median* of three or more genomes [11] and in phylogenetic reconstructions [19].

2.4   The General DCJ-Indel Model

In the DCJ-indel model the genomes can be unichromosomal or multichromosomal, linear or circular. We assign the cost of 1 to each DCJ operation and a positive cost w to each indel. Given two genomes A and B, the *DCJ-indel distance* of A and B, denoted by $d^{DCJ\text{-}id}(A, B)$, is the minimum cost of a sequence of DCJ and indel operations that sort A into B. If $w = 1$, the DCJ-indel distance corresponds exactly to the minimum number of steps required to sort A into B. For any positive $w \leq 1$, the DCJ-indel distance can be exactly computed in linear time [15,20].

Let S be a rearrangement scenario with DCJ and indel operations, we denote by $||S||$ the cost of S. By definition, $||S|| = n^{DCJ} + w(n^{ins} + n^{del})$, where n^{DCJ} is the number of DCJ operations and n^{ins} and n^{del} are, respectively, the number of insertions and deletions in S. If S is an optimal scenario sorting A into B, then $||S|| = d^{DCJ\text{-}id}(A, B)$.

Establishing the triangular inequality. The triangular inequality does not hold for the DCJ-indel distance, but a correction can be applied *a posteriori*, as proposed in [15,21]. It comprises summing to the distance a surcharge that depends on the number of unique markers. It has been shown that, given a positive constant $k = (w+1)/2$, for any $k' \geq k$ the triangular inequality holds for the function $m(A, B) = d^{DCJ\text{-}id}(A, B) + k'(|\mathcal{A}| + |\mathcal{B}|)$.

3   Restricted DCJ-Indel model

In the restricted DCJ-indel model the genomes are linear, unichromosomal or multichromosomal. We assign the cost of 1 to each DCJ operation and a positive cost $w \leq 1$ to each indel. Given two such genomes A and B, the *restricted DCJ-indel distance* of A and B, denoted by $d^{rDCJ\text{-}id}(A, B)$, is the minimum cost of a scenario of DCJ and indel operations that sort A into B, with the restriction that a circular excision has to be immediately followed by a circular reincorporation, forming an ER composition.

In this section we first solve an open problem from [16], before we present a simple algorithm to compute an optimal rearrangement scenario under the restricted DCJ-indel model.

3.1   Computing the Distance

Given two linear genomes A and B without duplicated markers, let S_1 be an optimal DCJ-indel scenario transforming A into B and let n^{DCJ}, n^{ins} and n^{del} be the number of DCJ operations, insertions and deletions in S_1, such that we have $d^{DCJ\text{-}id}(A, B) = ||S_1|| = n^{DCJ} + w(n^{ins} + n^{del})$.

As shown in [16], the scenario S_1 can be transformed into another optimal scenario S_2 of the same cost, so that S_2 starts with n^{ins} insertions, followed by n^{DCJ} DCJ operations, followed by n^{del} deletions. We can represent S_2 as follows:

$$S_2 = S_2^{ins} +\!\!+ S_2^{DCJ} +\!\!+ S_2^{del}$$

where S_2^{ins} is the prefix of S_2 with only insertions, S_2^{del} is the suffix of S_2 with only deletions, and S_2^{DCJ} is the substring of S_2 with DCJ operations. The symbol $+\!\!+$ denotes concatenation of rearrangement scenarios.

Let A' be the linear genome obtained after applying to A the insertions of S_2^{ins} and let B' be the linear genome obtained after applying to A' the DCJ operations of S_2^{DCJ}. Then the distance can be rewritten as

$$d^{DCJ\text{-}id}(A, B) = ||S_2^{ins}|| + d^{DCJ}(A', B') + ||S_2^{del}||.$$

Thus, A' and B' are two linear genomes with the same set of markers and DCJ distance $d^{DCJ}(A', B') = |S_2^{DCJ}|$, where $|S_2^{DCJ}|$ denotes the number of operations in S_2^{DCJ}. From [7, 13] we know that there exists a *restricted* DCJ scenario R of the same cost as S_2^{DCJ}, sorting A' into B'. Hence there also exists a restricted DCJ-indel sorting scenario S_3 transforming A into B:

$$S_3 = S_2^{ins} +\!\!+ R +\!\!+ S_2^{del}.$$

Clearly, S_3 has the same cost as S_2 and thus as S_1, being an optimal restricted DCJ-indel sorting scenario. These observations give rise to the following theorem:

Theorem 1. *Given two linear genomes A and B without duplicated markers, we have*

$$d^{rDCJ\text{-}id}(A, B) = d^{DCJ\text{-}id}(A, B).$$

Observe that Theorem 1 holds even if we assign the cost of 1 to each DCJ and a positive cost $w \le 1$ to each indel operation.

Complexity. For any positive indel cost $w \le 1$, the DCJ-indel distance can be computed in linear time [15, 20], and thus the same is true for the restricted DCJ-indel distance.

Establishing the triangular inequality. Obviously the correction proposed in [15, 21] to establish the triangular inequality for the DCJ-indel distance also holds for the restricted DCJ-indel distance.

Algorithm 1. Find a restricted DCJ-indel scenario sorting a linear genome A into a linear genome B

---

1. Compute an optimal DCJ-indel scenario S_1 sorting A into B using the algorithm from [15, 20].
2. Modify S_1 by moving the insertions up and the deletions down, as shown in [16], obtaining a scenario $S_2 = S_2^{ins} + S_2^{DCJ} + S_2^{del}$.
3. Use S_2^{ins} to transform A into a linear genome A'.
4. Use S_2^{DCJ} to transform A' into a linear genome B' (A' and B' have the same content $\mathcal{G}' = \mathcal{G} \cup \mathcal{A} \cup \mathcal{B}$).
5. Apply the restricted DCJ algorithm from [13] to obtain a restricted DCJ scenario R sorting A' into B'.
6. Concatenate the three parts to obtain the scenario $S_3 = S_2^{ins} + R + S_2^{del}$, that is a restricted DCJ-indel scenario sorting A into B.

3.2   Finding an Optimal Sorting Scenario

It can be easily seen that the procedure described in the previous subsection implies a simple algorithm for finding a restricted DCJ-indel scenario sorting a linear genome A into a linear genome B (Algorithm 1).

Complexity. In Algorithm 1, steps 1-4 and 6 can be implemented in linear time, while step 5 takes $O(n \log n)$ time, where $n = |\mathcal{G}'|$ is the number of markers in A', respectively B'. Thus, a restricted DCJ-indel sorting scenario can be computed in $O(n \log n)$ time.

Implementation. While an implementation of the restricted DCJ sorting is available in [22], to the best of our knowledge there exists no implementation of the general DCJ-indel sorting algorithm. Given such an implementation would then make it rather straightforward to also implement the restricted DCJ-indel sorting algorithm.

4   Conclusions and Perspectives

In this paper we have solved an open problem, showing that, even if the indel cost is distinct from and upper bounded by the DCJ cost, the restricted DCJ-indel distance is equal to the DCJ-indel distance, that can be computed in linear time. This allows the correction for establishing the triangular inequality in the DCJ-indel distance to be automatically extended to the restricted DCJ-indel distance.

We have also proposed an algorithm to generate an optimal restricted DCJ-indel sorting scenario in $O(n \log n)$ time. The most complicated parts of this algorithm are: (a) obtaining a general DCJ-indel sorting scenario between two genomes with unequal contents (step 1 of Algorithm 1) and (b) obtaining a restricted DCJ sorting scenario between genomes with equal contents (step 5 of Algorithm 1). An implementation of (b) is available in [22], but the implementation of (a) still has to be developed.

The inversion-indel distance. The inversion-indel is a related model that applies to unichromosomal (linear or circular) genomes only, and, instead of generic DCJ operations, allows only inversions of DNA segments, besides indels. An example is given in Figure 4. In [18] two algorithms were provided for this distance: an exact one for the case in which only one indel direction is allowed (i.e. when we have either only insertions or only deletions); and a heuristic for the symmetric case, in which both insertions and deletions are allowed. Recently, in a joint work with other authors [23], we proved that, for an important class of instances of the symmetric case, the inversion-indel distance equals the DCJ-indel distance. An exact solution for the general symmetric case remains an open problem.

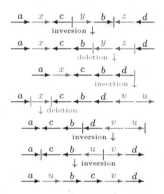

Fig. 4. An optimal sorting scenario in the inversion-indel model

The restricted DCJ-substitution distance. The DCJ-substitution is another related model that applies to linear genomes, unichromosomal or multichromosomal. In this model we have generic DCJ operations, but, instead of indels, more powerful operations are considered: *substitutions* allow blocks of contiguous markers to be replaced by other blocks of contiguous markers [24]. In other words, a deletion and a subsequent insertion that occur at the same position of the genome can be modeled as a substitution, counting together for one single step. In the DCJ-substitution model, indels are special cases of substitutions: if a block of markers is substituted by the empty string, we have a deletion; analogously, if the empty string is substituted by a block of markers, we have an insertion.

In the general DCJ-substitution model the results are very similar to the general DCJ-indel model. For a cost of 1 assigned to DCJ operations and any positive cost $w \leq 1$ assigned to substitutions, there is a formula to efficiently compute the distance [20, 24]. However, the general and the restricted DCJ-substitution distances are not the same, as we can see in the example given in Fig. 5. The restricted version of the DCJ-substitution distance is a complete open problem that we intend to study in the future.

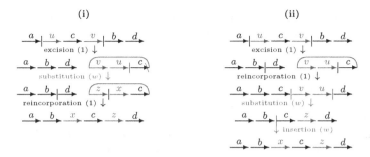

Fig. 5. (i) An optimal sorting scenario in the general DCJ-substitution model, with cost $2+w$. (ii) An optimal sorting scenario in the restricted DCJ-substitution model, with cost $2+2w$.

Acknowledgments. MDVB is funded by the Brazilian research agency CNPq grant PROMETRO 563087/10-2.

References

1. Hannenhalli, S., Pevzner, P.: Transforming men into mice (polynomial algorithm for genomic distance problem). In: Proc. of FOCS 1995, pp. 581–592 (1995)
2. Tesler, G.: Efficient algorithms for multichromosomal genome rearrangements. J. Comput. Syst. Sci. 65, 587–609 (2002)
3. Ozery-Flato, M., Shamir, R.: Two notes on genome rearrangement. J. Bioinf. Comput. Biol. 1, 71–94 (2003)
4. Jean, G., Nikolski, M.: Genome rearrangements: A correct algorithm for optimal capping. Inf. Process. Lett. 104, 14–20 (2007)
5. Bergeron, A., Mixtacki, J., Stoye, J.: A new linear time algorithm to compute the genomic distance via the double cut and join distance. Theor. Comput. Sci. 410, 5300–5316 (2009)
6. Erdős, P.L., Soukup, L., Stoye, J.: Balanced vertices in trees and a simpler algorithm to compute the genomic distance. Appl. Math. Lett. 24, 82–86 (2011)
7. Yancopoulos, S., Attie, O., Friedberg, R.: Efficient sorting of genomic permutations by translocation, inversion and block interchange. Bioinformatics 21, 3340–3346 (2005)
8. Bergeron, A., Mixtacki, J., Stoye, J.: A unifying view of genome rearrangements. In: Bücher, P., Moret, B.M.E. (eds.) WABI 2006. LNCS (LNBI), vol. 4175, pp. 163–173. Springer, Heidelberg (2006)
9. Adam, Z., Sankoff, D.: The ABCs of MGR with DCJ. Evol. Bioinform. Online 4, 69–74 (2008)
10. Mixtacki, J.: Genome halving under DCJ revisited. In: Hu, X., Wang, J. (eds.) COCOON 2008. LNCS, vol. 5092, pp. 276–286. Springer, Heidelberg (2008)
11. Tannier, E., Zheng, C., Sankoff, D.: Multichromosomal median and halving problems under different genomic distances. BMC Bioinformatics 10, 120 (2009)
12. Thomas, A., Varré, J.S., Ouangraoua, A.: Genome dedoubling by DCJ and reversal. BMC Bioinformatics 12(suppl. 19), S20 (2012)
13. Kováč, J., Warren, R., Braga, M.D.V., Stoye, J.: Restricted DCJ model (the problem of chromosome reincorporation). J. Comput. Biol. 18, 1231–1241 (2011)

14. Yancopoulos, S., Friedberg, R.: DCJ path formulation for genome transformations which include Insertions, Deletions, and Duplications. J. Comput. Biol. 16, 1311–1338 (2009)
15. Braga, M.D.V., Willing, E., Stoye, J.: Double cut and join with insertions and deletions. J. Comput. Biol. 18, 1167–1184 (2011)
16. da Silva, P.H., Machado, R., Dantas, S., Braga, M.D.V.: Restricted DCJ-indel model: sorting linear genomes with DCJ and indels. BMC Bioinformatics 13(suppl. 19), S14 (2012)
17. Braga, M.D.V.: An overview of genomic distances modeled with indels. In: Bonizzoni, P., Brattka, V., Löwe, B. (eds.) CiE 2013. LNCS, vol. 7921, pp. 22–31. Springer, Heidelberg (2013)
18. El-Mabrouk, N.: Sorting signed permutations by reversals and insertions/deletions of contiguous segments. J. Discr. Alg. 1, 105–122 (2001)
19. Swenson, K.M., Arndt, W., Tang, J., Moret, B.: Phylogenetic reconstruction from complete gene orders of whole genomes. In: Proc. of Asia-Pacific Bioinformatics Conf. Advances in Bioinformatics and Comp. Biology, vol. 6, pp. 241–250 (2008)
20. da Silva, P.H., Machado, R., Dantas, S., Braga, M.D.V.: DCJ-indel and DCJ-substitution distances with distinct operation costs. Alg. for Mol. Biol. 8, 21 (2013)
21. Braga, M.D.V., Machado, R., Ribeiro, L.C., Stoye, J.: On the weight of indels in genomic distances. BMC Bioinformatics 12(suppl. 9), S13 (2011)
22. Hilker, R., Sickinger, C., Pedersen, C., Stoye, J.: UniMoG - a unifying framework for genomic distance calculation and sorting based on DCJ. Bioinformatics 28, 2509–2511 (2012)
23. Willing, E., Zaccaria, S., Braga, M.D.V., Stoye, J.: On the inversion-indel distance. BMC Bioinformatics 14(suppl. 11), S3 (2013)
24. Braga, M.D.V., Machado, R., Ribeiro, L.C., Stoye, J.: Genomic distance under gene substitutions. BMC Bioinformatics 12(suppl. 9), S8 (2011)

MSA-GPU: Exact Multiple Sequence Alignment Using GPU

Daniel Sundfeld and Alba C.M.A. de Melo

Department of Computer Science, University of Brasília, Brasília, Brazil
{sund,alves}@unb.br

Abstract. In this paper, we propose and evaluate MSA-GPU, a solution to implement the exact Multiple Sequence Alignment algorithm in Graphics Processing Units (GPUs). In our solution, we use the Carrillo-Lipman upper and lower bounds to reduce the amount of computation. We propose a fine-grained strategy to explore the search space by using 2D projections. The results were obtained with a GTX 580 NVidia GPU comparing sets of 3 sequences (real and synthetic). We show that, for sequences with medium/low similarity, our GPU approach is able to outperform the MSA 2.0 CPU program, achieving a speedup of 8.6x.

1 Introduction

Bioinformatics is an interdisciplinary field that involves computer science, biology, mathematics and statistics [12]. One of its main goals is to analyze biological sequence data and genome content in order to obtain the function/structure of the sequences as well as evolutionary information.

Once a new biological sequence is discovered, its functional/structural characteristics must be established. In order to do that, the newly discovered sequence is compared against the sequences that compose genomic databases, in search of similarities. Sequence comparison is, therefore, one of the most basic operations in Bioinformatics. A sequence can be compared to another sequence (Pairwise Comparison), to a profile that describes a family of sequences (Sequence-Profile Comparison) or to a set of sequences (Multiple Sequence Alignment).

In a Multiple Sequence Alignment (MSA), similar characters among a set of k sequences ($k > 2$) are aligned together. Multiple Sequence Alignments are often used as a building block to solve important and complex problems in Molecular Biology, such as the identification of conserved motifs in a family of proteins, definition of phylogenetic relationships and 3D homology modeling, among others. In all these cases, the quality of the solutions relies heavily on the quality of the underlying multiple alignment. MSAs are often scored with the Sum-of-Pairs (SP) objective function and the exact SP MSA problem is known to be NP-complete [18]. Therefore, heuristic methods are often used to solve this problem, even when the number of sequences is small.

A great number of heuristic methods were proposed to tackle the Multiple Sequence Alignment problem. In a general way, they fall into two categories:

J.C. Setubal and N.F. Almeida (Eds.): BSB 2013, LNBI 8213, pp. 47–58, 2013.
© Springer International Publishing Switzerland 2013

progressive and iterative. A progressive MSA method initially generates all pairwise alignments and ranks them. The closest sequences are aligned first and then an MSA is built by adding the other sequences, in order of relevance. ClustalW [3] is an example of a progressive method. Iterative methods create an initial MSA of groups of sequences and then modify it, until a reasonable result is attained. DIALIGN [11] is an example of a deterministic iterative method. More recently, statistical methods that take into account evolutionary information such as Prank [7] and StatAlign [15] have also been proposed. This creates a great number of MSA heuristic tools, making it difficult to compare results and to determine the quality of a given MSA.

On the other hand, there do exist exact methods that are able to obtain the optimal Multiple Sequence Alignment [12]. The so-called naive exact method is a generalization of the exact algorithm based on dynamic programming that obtains optimal pairwise alignments [17]. It has time complexity $O(n^k)$, where n is the size of the sequences and k is the number of sequences.

Carrillo-Lipman [2] made an important contribution in the area of exact Multiple Sequence Alignment by showing that it is not necessary to explore the whole search space in order to obtain the optimal alignment. They showed that an heuristic alignment can be used to obtain an upper bound to the optimal alignment and all possible pairwise combinations can be used to obtain a lower bound. It is proven that the cells that fall outside these bounds do not contribute to the optimal alignment and, thus, these cells do not need to be calculated. Even with this, time complexity remains exponential.

Many efforts have been made to reduce the execution time of Multiple Sequence Alignment algorithms. Parallel versions were proposed to accelerate heuristic methods in clusters [5], [9], FPGAs (Field Programmable Gate Arrays) [16] and GPUs (Graphics Processing Units) [6], [1]. A few efforts were also made to implement exact Multiple Sequence Alignment algorithms in clusters [4] and FPGAs [8]. As far as we know, there are no implementations of the exact MSA in GPUs.

In this paper, we propose and evaluate MSA-GPU, a GPU solution to implement the exact Multiple Sequence Alignment algorithm. In our solution, we use the Carrillo-Lipman upper and lower bounds to reduce the amount of computation. We propose a fine-grained strategy to explore the search space by using 2D projections as in [8]. The results were obtained with a GTX 580 NVidia GPU comparing sets of 3 sequences (real and synthetic). We show that, for sequences with medium/low similarity, our GPU approach is able to outperform the MSA 2.0 CPU program, achieving a speedup of 8.6x.

The rest of this paper is organized as follows. We present an overview of the Multiple Sequence Alignment problem in Section 2. Section 3 discusses related work in the area of Exact MSA. In Section 4, the design of our GPU strategy for exact MSA is presented and the experimental results are shown in Section 5. Finally, we conclude the paper and give the future work directions in Section 6.

2   Multiple Sequence Alignment (MSA)

To compare two sequences, we search the best alignment between them, which amounts to place one sequence above the other making clear the correspondence between similar characters or substrings from the sequences [12].

A global Multiple Sequence Alignment (MSA) of $k > 2$ sequences $S = S_1, S_2, ..., S_k$ is obtained in such a way that spaces (gaps) are inserted into each of the n sequences so that the resulting sequences have the same length n. Then, the sequences are arranged in k rows of n columns each, so that each character or space of each sequence is in a unique column [12]. Figure 1 shows an example of one pairwise aligment and one MSA of 3 DNA sequences.

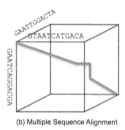

(a) Pairwise Alignment   (b) Multiple Sequence Alignment

Fig. 1. Pairwise alignment and MSA with 3 sequences. The gray line represents one possible alignment.

Usually, MSAs are scored with the Sum-of-Pairs (SP) function and the exact SP MSA problem is known to be NP-hard [18]. In SP, every pair of bases is scored with the pairwise scoring function and the final score is the addition of all these values [12]. For instance, considering that the punctuation for matches (similar characters), mismatches (different characters) and gaps are 0, $+1$ and $+1$, respectively, the score generated by pairwise comparison of sequences S_1 and S_2 (Figure 2) is $0 + 1 + 1 + 0 + 0 + 0 + 0 + 1 + 1 + 0 + 0 + 0 + 1 + 0 = 5$. The SP score of the MSA in this example is 16. When comparing proteins, a substitution matrix is used to score matches/mismatches. The most common substitution matrices are PAM and BLOSUM [12].

2.1   Heuristic Methods

Many heuristic methods have been proposed in the literature to solve the MSA problem. ClustalW [3] is a progressive heuristic method that aligns k sequences in three phases. In the first phase, all $(k * (k - 1)/2)$ pairwise alignments are computed and scores are obtained. These scores are used to generate a distance matrix that indicates the similarity between the sequences. In the second phase, a guide tree is generated from the distance matrix, using the neighbor-joining method. The guide tree is used in the third phase to progressively generate the MSA, starting with the most closely related sequences.

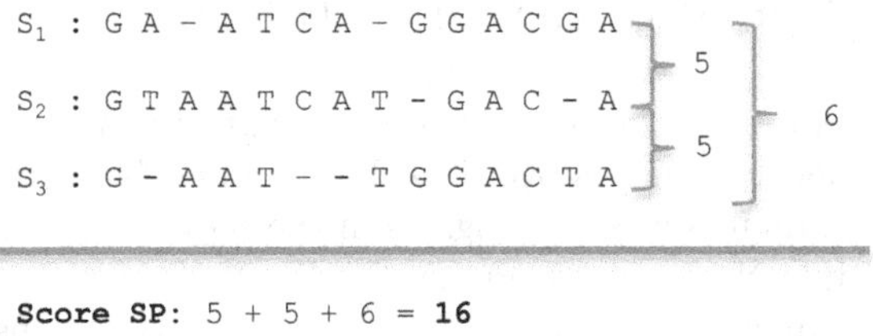

S_1 : G A - A T C A - G G A C G A
S_2 : G T A A T C A T - G A C - A
S_3 : G - A A T - - T G G A C T A

5
5
6

Score SP: 5 + 5 + 6 = **16**

Fig. 2. The Sum-of-Pairs scoring function

Like ClustalW, DIALIGN (DIagonal ALIGNment) [11] is an heuristic method for MSA. In order to align sequences, DIALIGN looks for ungapped fragments (or diagonals) and aligns them. Thus, in DIALIGN, an alignment is defined to be a chain of diagonals. The algorithm is executed in three phases. In the first phase, all DIALIGN pairwise alignments are computed, i.e., there are $k*(k-1)/2$ chains of diagonals, one for each pairwise alignment, where k is the number of sequences [10]. In the second phase, the diagonals that compose the pairwise alignments are sorted by their score and the degree of overlap with other diagonals. This sorted list is used to obtain an MSA with a greedy algorithm, generating alignment A. In the last phase, the alignment A is completed with an iterative procedure where the parts of the sequences that are not yet aligned with A are realigned by executing phase 2 again, in such a way that consistent non-aligned diagonals are included in A. This phase is repeated until no diagonal with a positive weight can be included in A.

In addition to ClustalW and DIALIGN, there are many other methods to calculate heuristic MSAs such as SAGA [14] and T-Coffee [13]. Even though the heuristic methods are able to provide good solutions, it is not guaranteed that the optimal MSA will be obtained.

2.2 Exact MSA

The optimal multiple sequence alignment among k sequences can be calculated by extending the exact dynamic programming algorithm for pairwise comparison [12]. Without loss of generality, assume that there are 3 sequences S_1, S_2 and S_3, of sizes n_1, n_2 and n_3, respectively.

The goal is to obtain the optimal score, which indicates the alignment distance (minimum number of insertions, deletions and substitutions). In order to do that, we calculate a 3-dimensional dynamic programming matrix D, where $D(i, j, k)$ is the optimal alignment of prefixes $S_1[1..i]$, $S_2[1..j]$ and $S_3[1..k]$.

The naive algorithm is depicted in Algorithm 1. There are three *for* loops (lines 1 to 3), one loop for each sequence. The scores for matches and mismatches are calculated in lines 4 to 6. After that (lines 7 to 13), seven values are calculated which correspond to the seven neighbor cells of $D(i, j, k)$. In line 14, $D(i, j, k)$ receives the minimum value of those calculated in lines 7 to 13. If there are k

Algorithm 1. Naive exact Multiple Sequence Alignment

1: **for** $i = 1 \rightarrow n_1$ **do**
2:    **for** $j = 1 \rightarrow n_2$ **do**
3:       **for** $k = 1 \rightarrow n_3$ **do**
4:          $c_{ij} = AssignMatchMismatchPunctuation(S_1(i), S_2(j))$;
5:          $c_{ik} = AssignMatchMismatchPunctuation(S_1(i), S_3(k))$;
6:          $c_{jk} = AssignMatchMismatchPunctuation(S_2(k), S_3(j))$;
7:          $d_1 = D(i - 1, j - 1, k - 1) + c_{ij} + c_{ik} + c_{jk}$
8:          $d_2 = D(i - 1, j - 1, k) + c_{ij} + gap$
9:          $d_3 = D(i - 1, j, k - 1) + c_{ik} + gap$
10:          $d_4 = D(i, j - 1, k - 1) + c_{jk} + gap$
11:          $d_5 = D(i - 1, j, k) + 2 * gap$
12:          $d_6 = D(i, j - 1, k) + 2 * gap$
13:          $d_7 = D(i, j, k - 1) + 2 * gap$
14:          $D(i, j, k) = Min[d_1, d_2, d_3, d_4, d_5, d_6, d_7]$
15:       **end for**
16:    **end for**
17: **end for**

---

sequences to be compared, there will be k loops in this algorithm and $2^k - 1$ cells will be used to calculate each cell of matrix D.

Carrillo-Lipman Bound. Carrillo-Lipman [2] showed that it is not necessary to explore the whole search space in order to obtain the optimal Multiple Sequence Alignment. They defined a lower and an upper bound which confine the region that contains the optimal alignment and thus restrict the area of the n-dimensional matrix to be calculated.

The lower (L) and upper (U) bounds are calculated as explained in the following paragraphs.

Equation 1 calculates the lower bound, based on the sum-of-pairs score. In this equation, $scale(S_i, S_j)$ is the weight of each pairwise aligment, usually choosen via an evolutionary tree of N sequences. $d(S_i, S_j)$ is the score of the optimal pairwise alignment of S_i and S_j .

$$L = \sum_{i<j} d(S_i, S_j) \cdot scale(S_i, S_j) \tag{1}$$

L is a lower bound for the following reason. Since $d(S_i, S_j)$ is the optimal score between sequences S_i and S_j, this is the lowest possible score for (S_i, S_j). Since the score of the Multiple Sequence Alignment is a sum-of-pairs, i.e., an addition of all pairwise scores, and the scores are non-negative, therefore the optimal Multiple Sequence Alignment score must be greater or equal to the sum of all optimal pairwise scores.

Consider that A^o is the optimal alignment, $c(A^o)$ is the score of the optimal alignment, $scale(S_i, S_j)$ is 1, and $c(A^o_{i,j})$ is the score of the pairwise alignment of i and j. The Carrillo-Lipman Bound is given by Inequation 2:

$$c(A_{i,j}^o) \leq d(S_i, S_j) + U - L \tag{2}$$

$U - L$ can be obtained by a heuristic alignment A^h. This alignment induces a score c in the sum-of-pairs i and j, and so $U - L$ is obtained by:

$$U - L = \sum_{i<j}[c(A_{i,j}^h) - d(S_i, S_j)] \tag{3}$$

Equation 3 shows that the lower and upper bounds have a value which is based on the projection of a heuristic alignment subtracted from the scores of the pairwise alignments. It is guaranteed that the cells that are outside those bounds do not contribute to the calculation of the optimal Multiple Sequence Alignment [2]. Frequently, the difference $U - L$ is called δ_{msa}.

3   Related Work

Even though there are many works in the literature that implement heuristic methods in GPUs [6], [1], FPGAs [16] and clusters [5], [9], there are very few works that implement exact MSA methods in high performace computing platforms.

Helal et al. [4] propose the use of a master-slave architecture to execute the exact MSA algorithm in a cluster. In order to reduce the search space, the authors use geometrical relations over hyper-diagonals and hyper-lattices. They were able to compare sets of 3, 4 and 5 sequences of small size in a cluster with 8 cores. The MSA comparison of 5 sequences took more than 2 days.

Masuno et al. [8] implemented the exact MSA with the Carrillo-Lipman bound in FPGA. In their proposal, the n-dimensional dynamic programming matrix is transversed in windows defined over 2D-projections, where i, j vary, whereas the other dimensions remain fixed. In order to implement the Carrillo-Lipman bound algorithm, a heuristic MSA is calculated in CPU and its score is transferred to the FPGA. Before calculating a window, the algorithm tests if it is inside the Carrillo-Lipman bound. If not, the window is not calculated. Two different circuits are proposed to calculate MSAs of 4 and 5 sequences, respectively. The computation of an MSA of 5 sequences took about 5 minutes in the FPGA.

4   Design of MSA-GPU

In order to execute the exact MSA algorithm, we opted to use GPUs since they provide massive SIMD (Single Instruction Multiple Data) parallelism for large-scale problems. For MSA-GPU, we designed two strategies, which are described in Sections 4.1 and 4.2.

4.1   Coarse-Grained Strategy

In the coarse-grained strategy, we used a multidimensional wavefront calculated by one GPU block and, inside the block, each GPU thread calculates one cell of the dynamic programming matrix D (Algorithm 1).

Figure 3 (coarse) illustrates the coarse-grained strategy, showing the tasks executed in the CPU and in the GPU. First, the user provides the sequences and the Carrillo-Lipman δ_{msa} (*delta_cl*) (Section 2.2). Then, the data structures in the GPU are initialized. After that, the kernels in GPU are executed for each multidimensional diagonal with n threads. The wavefront indexes are calculated in GPU and represent the coordinates of the cells that can be calculated in parallel. At the end of each kernel computation, the variable *bound_reached* is used to discard unnecessary multidimensional diagonals. At the very end of the computation, the optimal score is obtained and given as output to the user.

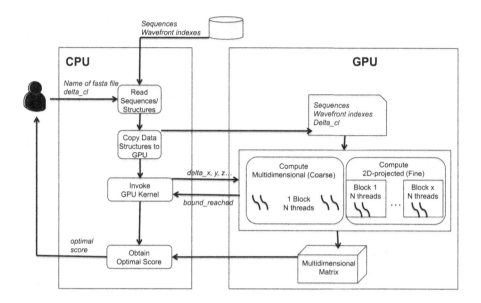

Fig. 3. Overview of the coarse-grained and fine-grained strategies

When using this coarse-grained strategy, we observed that the parallelism in the multidimensional wavefront grows a lot. But when the algorithm execute the last stages, some cells that might be calculated in paralell do not have the initial shape of a multidimensional wavefront. In this case, using the same wavefront shape would imply in the recalculation of some cells, reducing the throughput. The shape of the multidimensional wavefront processing is shown in Figure 4(a) and 4(b). In the left-corner picture in Figure 4(b) we can observe the effect of the cells without the wavefront shape.

4.2   Fine-Grained Strategy

With this strategy, we intended to augment the parallelism by using more than one GPU block. This was possible because we chose the wavefront indexes to

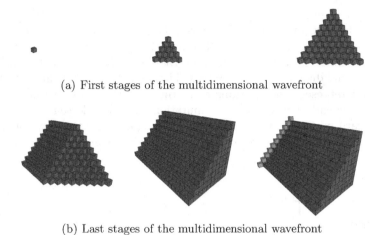

(a) First stages of the multidimensional wavefront

(b) Last stages of the multidimensional wavefront

Fig. 4. Multidimensional wavefront in the coarse-grained strategy

calculate the cells of the DP matrix using 2D projections. This led to a more regular dependency pattern and, thus, multiple blocks could be used.

The multi-block fine-grained strategy works as follows. Inside each block, the threads will calculate the same cell of the dynamic programming matrix. Therefore, lines 7 to 13 in Algorithm 1 will be parallelized, where each thread will calculate values d_1 to d_7 in parallel. Besides that, many cells that belong to the same projected diagonal will be calculated in parallel by different blocks. Figure 3 (fine) illustrates this strategy.

In this strategy, we traverse the search space using 2D projections, as shown in Figure 5(a) and 5(b). In this figure, it can be seen that a more regular pattern is obtained.

5   Experimental Results

The strategies proposed in Section 4 were implemented in CUDA C, using the CUDA toolkit 4.1.21. The results were obtained with the NVidia GTX 580 GPU (512 cores and 1.5 GB RAM). This GPU was connected to an Intel Core i5 host machine, with 6GB RAM.

In our tests, we used real and synthetic sequences (Table 1). The real sequences were obtained from the PFAM (*pfam.sanger.ac.uk*) and Balibase 2.0 (*bips.u-strasbg.fr/fr/Products/Databases/BAliBASE2/*) databases. The synthetic sequences were constructed in order to reproduce easy, medium, hard and very hard MSA patterns.

First, we compared the sequences in Table 1 with the coarse (Section 4.1) and the fine granularity (Section 4.2) approaches. In the last case, one block and multiple blocks were used. The results are shown in Table 2. In this table, we can see that, when the sequences are small, the fine-grained strategy achieves better

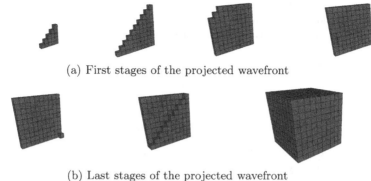

(a) First stages of the projected wavefront

(b) Last stages of the projected wavefront

Fig. 5. 2D-projected wavefront in the fine-grained strategy

Table 1. Sequences used in the Tests

Name	Database	Reference	Sizes		
Seq15	PFAM	PF10550	15	15	14
Seq22	PFAM	PF08095	22	22	22
Seq42	PFAM	PF03855	41	44	42
Seq59	PFAM	PF08184	59	59	59
Seq110	PFAM	PF11513	106	111	110
Seq122	PFAM	PF06453	122	122	122
Seq143	PFAM	PF09155	143	143	143
Seq162	PFAM	PF03426	158	158	162
Seq373	Balibase2	1pedA	350	326	373
Seq446	Balibase2	1ad3	423	441	446
Seq416	Synthetic	synthetic_easy	231	416	363
Seq417.1	Synthetic	synthetic_medium	231	447	363
Seq417.2	Synthetic	synthetic_hard	231	447	423
Seq453	Synthetic	synthetic_veryhard	231	446	453

results than the coarse-grained strategy, even with only one block. When we augment the sizes of the sequences, the coarse-grained strategy surpasses the fine-grained strategy (one block) since the reduced number of threads is insufficient to deal with the parallelism. For all the sequences compared, the fine-grained strategy (multiple blocks) was able to achieve the best execution times, with a great improvement over the other approaches. For instance, when comparing sequences *Seq446*, the execution time was reduced from 8min55s (coarse-grained) to 6.62s (fine-grained multi-block).

We also compared the execution times of the fine-grained multi-block GPU strategy with the MSA 2.0 CPU program. This program is publicly available at *www.ncbi.nim.nih.gov/CBBresearch/Schaffer/msa.html*, runs in CPU and is often used to obtain exact multiple sequence alignments. The execution times in our host machine (one core) and in the GPU are shown in Table 3. In this table,

Table 2. Comparison Between the Coarse-Grained (C-Grain) with 990 Threads, Fine-Grained One Block (F-Grain-1B) with 7 Threads, and the Fine-Grained Multiple Blocks (F-Grain-MB) with Variable Number of Threads

Name	C-Grain	F-Grain-1B		F-Grain-MB
Seq15	1.09s	0.12s	0.05s	7 to 105 threads
Seq22	1.34s	0.27s	0.07s	7 to 154 threads
Seq42	3.61s	1.68s	0.10s	7 to 308 threads
Seq59	4.60s	4.28s	0.14s	7 to 413 threads
Seq110	16.51s	27.25s	0.35s	7 to 777 threads
Seq122	26.05s	37.71s	0.46s	7 to 854 threads
Seq143	29.84s	1min04s	0.47s	7 to 1001 threads
Seq162	41.07s	1min25s	0.74s	7 to 1106 threads
Seq373	5min34s	14min23s	4.04s	7 to 2282 threads
Seq416	4min16s	16min12s	3.69s	7 to 2912 threads
Seq446	8min55s	29min30s	6.62s	7 to 3087 threads
Seq447.2	5min11s	>30 min	6.49s	7 to 3129 threads

Table 3. Execution Times for the MSA 2.0 CPU program and the MSA-GPU fine-grained multi-block GPU strategy using Full Space Search (FSS) and reducing the Search Space with Carrillo-Lipman (CL)

Name	δ_{msa}	MSA 2.0 CPU (s)	MSA-GPU (FSS) (s)	MSA-GPU (CL) (s)
Seq15	15	0.001	0.051	0.051
Seq22	15	0.001	0.068	0.056
Seq42	15	0.002	0.102	0.096
Seq59	15	0.002	0.140	0.126
Seq110	15	0.005	0.353	0.313
Seq122	15	0.006	0.461	0.435
Seq143	15	0.007	0.473	0.423
Seq162	15	0.010	0.743	0.650
Seq373	137	0.180	4.041	3.324
Seq446	29	0.140	6.624	6.203
Seq416	502	1.652	3.693	2.611
Seq417.1	185	23.507	3.837	3.011
Seq417.2	484	28.711	6.478	4.764
Seq 453	387	31.078	6.948	3.612

the second column presents the δ_{msa} parameter given to the MSA-GPU program to use the Carrillo-Lipman bound. For the MSA 2.0 CPU program, the default parameters were used, whenever possible. For sequences *Seq447.1, Seq447.2* and *Seq453*, the MSA 2.0 program was not able to retrieve the score/alignment with the default parameters. Therefore, we had to augment the δ_{msa} parameter to 100, 100 and 1000, respectively. The third, forth and fifth columns show, respectively, the execution times for the MSA 2.0 CPU program, the MSA-GPU fine-grained multi-block strategy (Full Space Search) and the MSA-GPU fine-grained multi-block strategy (Carrillo-Lipman).

We can see that, for sequences *Seq15* to *Seq446*, the MSA 2.0 CPU program is able to execute very quickly, with much better execution times than the GPU program. This happens because the sequences have high similarity and, for this reason, the CPU program is able to prune efficiently the search space. Even though our GPU program also prunes the search space, these sequence sets do not have enough parallelism to surpass the CPU implementation.

Sequences *Seq447.1, Seq447.2* and *Seq453* are more complex cases. Therefore, the MSA 2.0 program was not able to execute with the default parameters and the δ_{msa} parameter was augemented. Augmenting the δ_{msa} parameter reduces the area pruned by the Carrillo-Lipman bound , thus augmenting the execution time. For these cases, the GPU program presents a speedup of 7.8x for the sequence set *Seq447.1* (medium similarity), 6.02x for *Seq447.2* (low similarity) and 8.60x for *Seq453* (very low similarity).

6 Conclusion and Future Work

In this paper, we proposed and evaluated MSA-GPU, a parallel tool that is able to calculate exact MSAs in GPUs. We proposed coarse-grained and fine-grained mechanisms to express the parallelism, with different strategies to compute the dynamic programming matrix (multidimensional and 2D-projected wavefronts).

The results obtained with real and synthetic sequence sets composed of 3 sequences show that the fine-grained multiblock strategy achieves better execution times than the coarse-grained strategy when the sequences have a reasonable size. When comparing the fine-grained multi-block MSA-GPU with the MSA 2.0 CPU program, we observed that the CPU program has very low execution times, when the sequences are similar. For sequences with medium/low similarity, the execution time augments a lot and, in these cases, MSA-GPU outperforms MSA 2.0, being able to reduce the execution time considerably.

As future work, we intend to extend MSA-GPU to compare up to 7 sequences. We also intend to investigate alternative bounds to the Carrillo-Lipman bound that guarantee that the optimal result will be produced by calculating a smaller area in the n-dimensional dynamic programming matrix.

References

1. Blazewicz, J., Frohmberg, W., Kierzynka, M., Wojciechowski, P.: G-MSA - A GPU-based, fast and accurate algorithm for multiple sequence alignment. Journal of Parallel and Distributed Computing 73(1), 32–41 (2013)
2. Carrillo, H., Lipman, D.: The Multiple Sequence Alignment Problem. SIAM Journal of Applied Math. 48, 1073–1082 (1988)
3. Higgins, D.G., Thompson, J.D., Gibson, T.J.: ClustalW: improving the sensitivity of progressive multiple sequence alignment through sequence weighting, position-specific gap penalties and weight matrix. Nucleic Acids Research 22, 4673–4680 (1994)
4. Helal, M., Mullin, L.R., Potter, J., Sintchenko, V.: Search Space Reduction Technique for Distributed Multiple Sequence Alignment. In: NPC, pp. 219–226 (2009)

5. Li, K.B.: ClustalW-MPI: ClustalW analysis using distributed and parallel computing. Bioinformatics 19(12), 1585–1586 (2003)

6. Liu, Y., Schmidt, B., Maskell, D.L.: MSA-CUDA: Multiple Sequence Alignment on Graphics Processing Units with CUDA. In: ASAP, pp. 121–128 (2009)

7. Loytynoja, A., Goldman, N.: Phylogeny-aware gap placement prevents errors in sequence alignment and evolutionary analysis. Science 320, 1632–1635 (2008)

8. Masuno, S., Maruyama, T., Yamaguchi, Y., Konagaya, A.: An FPGA Implementation of Multiple Sequence Alignment Based on Carrillo-Lipman Method. In: Field Programmable Logic and Applications, pp. 489–492 (2007)

9. Macedo, E.A., Melo, A.C.M.A., Pfitscher, G.H., Boukerche, A.: Multiple biological sequence alignment in heterogeneous multicore clusters with user-selectable task allocation policies. The Journal of Supercomputing 63(3), 740–756 (2013)

10. Morgenstern, B., Dress, A., Werner, T.: Multiple DNA and protein sequence alignment based on segment-to-segment comparison. PNAS, USA, 12098–12103 (1996)

11. Morgenstern, B., Frech, K., Dress, A., Werner, T.: DIALIGN: Finding local similarities by multiple sequence alignment. Bioinformatics 14(3), 290–294 (1998)

12. Mount, D.W.: Bioinformatics: sequence and genome analysis. Cold Spring Harbor Laboratory Press (2004)

13. Notredame, C.: T-Coffee: a novel method for fast and accurate multiple sequence alignment. Journal of Molecular Biology 302, 205–217 (2000)

14. Notredame, C., Higgins, D.G.: SAGA: sequence alignment by genetic algorithm. Nucleic Acids Research 24, 1515–1524 (1996)

15. Novak, A., Miklos, I., Lyngso, R., Hein, J.: StatAlign: an extendable software package for joint Bayesian estimation of alignments and evolutionary trees. Bioinformatics 24(20), 2403–2404 (2008)

16. Oliver, T.F., Schmidt, B., Nathan, D., Clemens, R., Maskell, D.L.: Using reconfigurable hardware to accelerate multiple sequence alignment with ClustalW. Bioinformatics 21(16), 3431–3432 (2005)

17. Smith, T.F., Waterman, M.S.: Identification of common molecular subsequences. Journal of Molecular Biology 147(1), 195–197 (1981)

18. Wang, L., Jiang, T.: On the complexity of multiple sequence alignment. Journal of Computational Biology 4, 337–348 (1994)

MultiSETTER - Multiple RNA Structure Similarity Algorithm

David Hoksza[1], Peter Szépe[1], and Daniel Svozil[2]

[1] Charles University in Prague, FMP, Department of Software Engineering,
Malostranske nam. 25, 118 00 Prague, Czech Republic
hoksza@ksi.mff.cuni.cz
http://siret.cz/hoksza
[2] Institute of Chemical Technology, Laboratory of Informatics and Chemistry,
Technicka 5, 166 28 Prague, Czech Republic

Abstract. Recent advances in RNA research and a steady growth of available RNA structures call for bioinformatics methods for handling RNA structural data. Recently, we have introduced SETTER — a fast and accurate method for RNA pairwise structural alignment. In the present contribution we describe its extension for multiple RNA structure alignment called MultiSETTER. It combines SETTER's decomposition of RNA structures into disjoint fragments with a well known multiple sequence alignment algorithm ClustalW adapted for the structural alignment. We demonstrate the validity of our approach on the task of automatic classification of RNA structures.

Keywords: structural bioinformatics, RNA tertiary structure, structural similarity.

1 Motivation

The biological research in recent twenty years revealed an appreciable role of RNA in many important biological processes [1, 2]. A thorough understanding of various roles of RNA, such as its importance in gene expression regulation, significantly expands our appreciation of genome-related processes, and the RNA research field has significantly evolved in the last decade. As a consequence, a number of elucidated 3D RNA structures is steadily growing, a trend which can be expected to be even pronounced in the following years. With a growing number of RNA structures also grows a need for the development of domain specific algorithms for searching in RNA databases. Recently, we introduced an RNA structure pairwise alignment algorithm SETTER [3], and a web server [4] utilizing SETTER for searching in an 3D RNA structure database. In this paper, we propose an extension of this approach allowing to assess a similarity to a group of RNA structures based on multiple structure alignment. Such an approach can be used, e.g., for an automatic classification of RNA structures. While several multiple RNA sequence alignment algorithms are available [5–7], we are not aware of any algorithm capable of performing multiple 3D structural alignment.

J.C. Setubal and N.F. Almeida (Eds.): BSB 2013, LNBI 8213, pp. 59–70, 2013.

However, there exist algorithms performing multiple alignment based on the secondary structure [8, 9] or algorithms which project structural features into sequence and carry out multiple sequence alignment over this projection [10].

2   Pairwise Structural Alignment Using SETTER

Similarly to protein structure, RNA structure can be described at four levels of detail. Primary structure is represented by the RNA's linear sequence of nucleotides. Unlike DNA, RNA comes single stranded, and is free to form a multitude of intramolecular hydrogen bonds. Complex base-pairing patterns consisting of double helices interconnected by various types of loops [11] represent the second level of the structural hierarchy – secondary structure. Tertiary structure is formed by the overall arrangement and packing of double helices [12]. Finally, in case of multiple-chain RNA structures the mutual arrangement of tertiary structures of individual chains forms quaternary RNA structure.

SETTER utilizes base-pairing information to divide an RNA structure into basic alignment elements called generalized secondary structure units (GSSUs) [3]. A GSSU can be viewed as a simplified version of a secondary structural motif. Each GSSU consists of three parts – a stem, a neck and a loop. RNA structure is formed by non overlapping GSSUs which are identified along its sequence (Fig. 1).

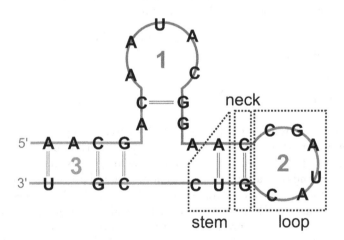

Fig. 1. Three GSSUs extracted from an RNA structure. The sequence starts at the 5′ end. Borders between individual GSSUs are indicated by the dashed lines and the numbers show the order of the GSSU generation.

To compare a pair of RNA structures R_1 and R_2 each GSSU from R_1 is aligned with each GSSU from R_2. To align a pair of GSSUs, three key residues from each GSSU are superimposed using the Kabsch algorithm [13]. The key residues are formed from two pairs of neck residues and one pair of loop residues chosen so that it yields best superposition. The quality of a superposition is assessed by the S-score that represents the similarity of a GSSU pair. If structures consist of more than one GSSU each, a pair of GSSUs G_1^i, G_2^j with the lowest S-score drives the superposition of R_1 and R_2 structures. To superimpose R_1 and R_2 a translation vector and a rotation matrix defining the superposition of G_1^i, G_2^j is utilized. The quality of a structural alignment is evaluated by aggregating S-distances of all superposed GSSUs into a score $\overline{S}$. For further details regarding the SETTER algorithm we refer to the original publication [3].

3   Multiple Structural Alignment Using MultiSETTER

The MultiSETTER approach represents a SETTER's extension for multiple structural alignment. Our approach adopts principles used in ClustalW [14] algorithm for multiple sequence alignment of proteins. The ClustalW algorithm can be summarized in the following three steps:

- Perform all possible alignments between each pair of sequences, and generate a distance matrix.
- From the distance matrix construct a dendrogram called "guide tree" using the neighbor-joining method [15].
- Construct a multiple sequence alignment by aligning the sequences in the order defined by the guide tree.

In MultiSETTER, we transformed sequence-specific parts of ClustalW into their structure counterparts. Hence, the MultiSETTER algorithm works as follows:

- Each pair of RNA structures is aligned by SETTER, and a distance matrix is constructed from the $\overline{S}$ distances of each RNA pairwise alignment.
- A guide tree is calculated from the distance matrix.
- The two most closely related structures are aligned first resulting in a so-called *average RNA structure* that blends structural characteristics of both input RNA structures. The alignment then progressively continues, generating more and more average structures, until the root of the guided tree is reached. The root of the tree corresponds to the final multiple structure alignment.

Two RNA structures are merged (averaged) utilizing their decomposition into individual GSSUs. Merging whole structures thus takes place independently on the level of GSSUs and these independently merged pieces of the structures (GSSUs) are then aggregated into the final average RNA structure. The result of the aggregation is a newly formed RNA structure which is equivalent to any other non-artificial RNA structure.

The complexity of the pairwise RNA comparison using SETTER is $O(n^2)$ for single GSSU structures where n is the number of nucleotides. If the larger structure contains m GSSUs each having with at most n nucleotides then the complexity is $O(m^2 \times n^2)$. This looks as a high value but one has to take into account that the separation into GSSUs keeps the GSSUs relatively small thus efficiently limiting both $O(m^2)$ and $O(n^2)$ parts. Moreover, additional heuristics are applied to further decrease the complexity (see [3] for details). When multiple comparison takes place, first the all-to-all RNA pairwise distance matrix is computed and then the merging step is carried out. If there are k structures to be aligned, the complexity of the all-to-all comparison is $O(k^2 \times m^2 \times n^2)$. Since the complexity of the merging step is linear (see the following section) the last mentioned complexity is also the complexity of MultiSETTER. How this complexity reflects itself in the runtimes shows section 4.2.

In the following sections we give a detailed description of the merging process.

3.1   RNA Merging

The average structure is constructed from two RNA molecules decomposed into GSSUs. To merge two structures a list of GSSU pairs G formed during the pairwise alignment, a translation vector and a rotation matrix are utilized. First a so-called *lead structure L* is identified. The lead structure is that structure from the pair to be aligned with the lower sum of distances to the rest of structures. This prevents the multiple alignment to be disrupted by outliers. In the following step the algorithm takes GSSUs from L and identifies its corresponding GSSU in G thus forming a pairing on the level of GSSUs. These GSSUs are then merged (see section 3.2) into new $GSSU_M$. This process is described by the following code where $GSSU_L$ is the GSSU from the *lead structure L*, and $GSSU_O$ is the GSSU from the second structure.

$GSSU_M \leftarrow GSSU_L$
if S($GSSU_L$, $GSSU_O$) < meanDistance * param1 **then**
   $GSSU_M \leftarrow$ merge($GSSU_L$, $GSSU_O$);
   if S($GSSU_M$, $GSSU_L$) + S($GSSU_M$, $GSSU_O$) < S($GSSU_L$, $GSSU_O$) * param2
   then
      $GSSU_M \leftarrow GSSU_L$
   end if
end if

The variable *meanDistance* represents the average distance between all GSSUs in the input set. If the distance between two GSSUs diverges significantly from the average the merging is bypassed and the GSSU from the lead structure is used as $GSSU_M$. Similarly, the merged GSSU is not accepted if its distance to the $GSSU_L$ is greater than the distance between the two original GSSUs. The parameters *param1* and *param2* decrease the influence of outliers in the alignment. The higher the value of *param1*, the easier it is for a pair of GSSUs to pass the condition and thus be merged. The same reasoning is valid for *param2*.

3.2   GSSU Merging

The GSSU merging algorithm merges loops and stems separately, with the neck being considered as a part of the stem (see Fig. 2a).

Algorithm 1. Merge stem *(merges r_j stem of $GSSU_L$ and $GSSU_O$ and outputs $GSSU_M$)*

```
 1: translationVector ← (0,0,0);
 2: translationDir ← 1;
 3: stem1 ← GSSU_L.stem; stem2 ← GSSU_O.stem;
 4: if GSSU_L.stem.Length() < GSSU_O.stem.Length() then
 5:    stem1 ← GSSU_O.stem; stem2 ← GSSU_L.stem;
 6: end if
 7: if stem2.Length = 0 then
 8:    GSSU_M.stem.side[r_j] ← stem1.side[r_j];
 9:    return GSSU_M;
10: end if
11: step ← stem1.Length() / stem2.Length();
12: for i = stem1.Length() downto 0 do
13:    residue1 ← stem1[i].r_j; residue2 ← stem2[step * i].r_j;
14:       /* Select a residue based on the merging rules. */
15:    pos ← i;
16:    mergedResidue ← SelectResidue(residue1, residue2, pos);
17:    if residue1 ! = NULL AND residue2 ! = NULL then
18:       translationVector ← CalculateTranslation(residue1, residue2, merge-
          dResidue);
19:       translationSource ← mergedResidue.Template;
20:    else
21:          /* One of the residues not present ⇒ translation vector from the
             previous iteration used. */
22:       if mergedResidue.Template = translationSource then
23:          translationDir ← 1;
24:       else
25:          translationDir ← -1;
26:       end if
27:       TranslateAtoms(mergedResidue.atoms, translationDir * translationVector);
28:       mergedGSSU.atoms.append(mergedResidue.atoms);
29:       mergedGSSU.stem[i].side[r_j] ← mergedResidue;
30:    end if
31: end for
```

Stem Merging. Stem merging follows the Algorithm 1. The algorithm input consists of the information about which side ($r1$ or $r2$ — see Fig. 2a) of which GSSUs to merge. The algorithm moves stepwise over the stem and merges residues in every step. Each time, one residue is selected as a template which is positioned by applying a translation vector (see the next paragraph). To select a template residue (function *SelectResidue*, line 16) we implemented two approaches. In the first approach, referred to as *Leading*, residues from the $GSSU_L$ (if present) are favored. In the second approach, called *Alternating*, we use a modulo 2 function (see Fig. 2b).

It is important to position the atoms of the new (i.e., merged) residue correctly. A translation vector is calculated by the *CalculateTranslation* function, line 18. This function is called only if residues in both stems are present. The resulting vector translates the template residue to the midpoint between the input residues. If the template residue comes from the same side as the selected residue from a previous step when both residues were present, the translation vector is used as is. Otherwise, it is multiplied by -1 to reverse its direction.

Loop Merging. The method for loop merging is the same as the stem merging algorithm. The only difference is that the *SelectResidue* function is no longer needed as at each loop's position there exist both residues. Which of them will be used depends on the strategy used in function *SelectResidue*.

4   Experimental Evaluation

We used MultiSETTER to automatically classify RNA structures for which we used the SCOR classification. SCOR [16, 17] (Structural classification of RNA) is a human curated database of 3D RNA structures hierarchically organized based on their function and tertiary interactions. At the lowest level of the SCOR hierarchy functionally similar structures form so-called "families". From SCOR database we extracted 14 families (each with at least four structures) of various sizes and intrinsic diversity containing 101 structures in total. SCOR names of the families, PDB codes of structures and their chain IDs are given in the following list (the last number shows the average number of nucleotides per structure).

1. **tRNA (Phe):** 1EHZ-A, 1EVV-A, 1TN2-A, 1TRA-A, 4TNA-A, 4TRA-A, 6TNA-A *(76)*
2. **tRNA (Gln):** 1C0A-B, 1EFW-C, 1IL2-C, 1ASY-R, 1ASZ-R *(75)*
3. **tRNA (Asp):** 1EUY-B, 1EXD-B, 1GTR-B, 1GTS-B, 1O0C-B, 1QRS-B, 1QTQ-B *(73)*
4. **Synthetic:** 1J4Y-A, 1KKA-A, 1LUU-A, 1LUX-A *(16)*
5. **Zymomonas mobilis:** 1Q2R-E, 1Q2R-F, 1Q2S-E, 1Q2S-F *(19)*
6. **tRNA (Lys):** 1BZ2-A, 1BZ3-A, 1BZT-A, 1BZU-A, 1FEQ-A, 1FL8-A *(16)*
7. **Haloarcula marismortui:** 1JJ2-9, 1K73-B, 1K8A-B, 1K9M-B, 1KC8-B, 1KD1-B, 1KQS-9, 1M1K-B, 1M90-B, 1N8R-B, 1NJI-B, 1Q7Y-B, 1Q81-B, 1Q82-B, 1Q86-B, 1QVF-9, 1QVG-9, 1S72-9 *(121)*
8. **P5 stem loop:** 1C0O-A, 1EOR-A, 1F9L-A, 1GUC-A *(16)*
9. **Hammerhead ribozyme:** 1NYI-A, 1Q29-A, 1HMH-A, 1MME-A, 299D-A, 359D-A, 379D-A, 488D-A *(18)*
10. **MS2 phage coat protein binding stem-loop:** 1AQ3-A, 1D0T-A, 1D0U-A, 1DZS-A, 1GKV-R, 1GKW-R, 1H8J-R, 1KUO-R, 1ZDH-A, 1ZDI-A, 1ZDJ-A, 1ZDK-A *(16)*
11. **HIV-1 TAR RNA:** 1ANR-A, 1ARJ-A, 1QD3-A, 397D-A, 1AKX-A, 1LVJ-A, 1UTS-B, 1UUD-B, 1UUI-A *(27)*
12. **HIB-1 Rev response element Rev binding site:** 1CSL-A, 1DUQ-A, 1EBQ-A, 1EBR-A, 1EBS-A, 1ETF-A, 1I9F-A *(25)*
13. **HIV-1 psi RNA stem loop SL1:** 1M5L-A, 1N8X-A, 1JTJ-A, 1JU1-A, 1F6U-A, 1OSW-A *(26)*
14. **RNA double strand, bound to protein:** 1A34-A, 2BBV-A, 1RC7-A, 1DI2-A, 1N35-A *(8)*

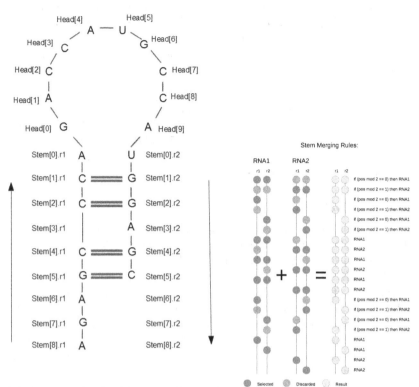

(a) An example of a GSSU consisting of a stem and a loop. Residues *Stem[0].r1* and *Stem[0].r2* form a neck, however, the neck is considered to be a part of the loop in the GSSU merging algorithm.

(b) *Alternating* schema of merging two stems. *Pos* is the placeholder of the position of a pair in a stem.

Fig. 2. Stem merging

We assigned MultiSETTER's accuracy using the leave-one-out classification of all 101 structures. We employed the following protocol:

– Pick one of the 101 structures and set it as the query structure RNA_Q.
– Remaining 100 structure form a database with 14 structural classes.
– Compute an average structure for each of the 14 classes.
– Compute distances between the RNA_Q and each of the 14 average structures.
– Assign RNA_Q into the family with closest average structure.

We compared the multiple alignment based classification obtained by MultiSETTER with the SETTER's pairwise alignment based classification. To classify a query structure with SETTER, we computed its distance to the remaining 100 structures. We assigned a class of the query structure using two approaches:

- *Best* — We sorted the database of 100 structures in a decreasing order of their distance to the query structure. We assigned the query structure into the class of the closest database hit.
- *Mean* — For every of the 14 families we computed the average distance between the query structure and the family members. The resulting classification was the class with the lowest average distance.

We ran SETTER with its default parameters [4] in all experiments. Multi-SETTER's parameters *param1* and *param2* were set to 0.9 and 2.0, respectively. For testing we used a machine with an Intel Core i7-2620M processor with 2 cores and 4 threads, 4 Gigabytes of RAM and hard drive with 7200 rpm; running on 64 bit version of Microsoft Windows 7.

4.1 Accuracy Evaluation

The accuracy evaluation of RNA structural classification by MultiSETTER and SETTER methods is summarized in Tab. 1. *Alternating* and *Leading* setups (see section 3.2) were used in MultiSETTER, and *Best* and *Mean* setups (see section 4) in SETTER. Each row in Tab. 1 contains the percentage of 101 queries for which the class of the query structure ended at that position (column "Position"). Thus, Position 2 means that the first structure belonging to same structural class as the query is found at the second position in the list of structures sorted by their distances to the query. It then follows that the structure at the first position in the same example, though it is closer to the query, is annotated with a wrong classification.

Tab. 1 shows that most accurate results are obtained by pairwise structural alignment performed by SETTER with the *Best* setup (i.e., most similar structure is used to assign a functional class). Good results of the *Best* setup are the consequence of not relying on average structures which can be easily influenced by any outlying structures. Although MultiSETTER was designed with the outliers' possible influence in mind, still, the average structure can be influenced by outlier(s), and in such a case a classification based on the multiple alignment might not be the best solution. However, if we compare MultiSETTER (both *Alternating* and *Leading* setups) with the SETTER's *Mean* setup also employing a structure averaging, we can see that multiple alignment yields much better results. MultiSETTER is therefore better than averaging over individual pairwise alignments.

The conclusions from the previous paragraph are consistent with a visual inspection of the alignments. Fig. 3 shows average positions of correct classifications broken down to individual families. For example, if there are only two structures in the family F, first structure ends at second, and second structure at fourth position during the classification, the average position of the family F is three. Fig. 3 demonstrates that MultiSETTER typically outperforms SETTER-Mean and in some cases also SETTER-Best approaches. In cases when Multi-SETTER is outperformed by SETTER we can usually track down the problem to one structure which is the outlier with respect to the rest in the family. For example, in family 9 the outlying structure is 1HMH-A.

Table 1. Comparison of MultiSETTER and SETTER

Position	MultiSETTER		SETTER	
	Alternating (%)	Leading (%)	Best (%)	Mean (%)
1	79.38	78.35	86.60	70.10
2	7.22	2.06	2.06	4.12
3	3.09	5.15	5.15	2.06
4	1.03	3.09	0.00	6.19
5	1.03	2.06	1.03	6.19
6	1.03	1.03	0.00	2.06
7	1.03	2.06	2.06	0.00
8	1.03	0.00	0.00	2.06
9	0.00	0.00	2.06	1.03
10	2.06	2.06	0.00	1.03
11	2.06	3.09	0.00	1.03
12	1.03	0.00	0.00	1.03
13	0.00	0.00	1.03	3.09
14	0.00	1.03	0.00	0.00

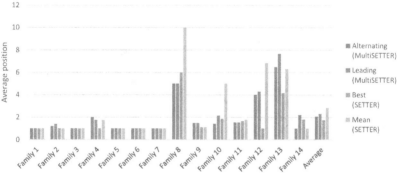

Fig. 3. Average positions of correct classifications of all the SETTER's and MultiSET-TER's setups. A lower value stands for better classification performance.

Finally, we visually verified that multiple alignment classifies correctly structures sharing a common fold. Fig. 4a — Fig. 4d show multiple alignments of four families differing by their level of structural homology. Family 1 (the numbering corresponds to the list at the beginning of section 4) in Fig. 4a is highly homologous, while family 12 at Fig. 4c is rather diverse. Therefore, family 12 is best classified by a pairwise alignment with the *Best* setup (see Fig. 3). In addition, in Fig. 4d we demonstrate that MultiSETTER is also able to align dataset consisting of large structures. In this specific case we used seven ribosomal subunits — PDB IDs 1JJ2-0, 1K8-A, 1K9M-A, 1M1K-A, 1NJP-0, 1NKW-0, 1S72-0.

Table 2. Time in seconds (s) needed to create the average structures for individual families. Single-threaded and four-threaded runtimes are reported.

Family	1	2	3	4	5	6	7	8	9	10	11	12	13	14
1 thread	7.9	3.4	8.0	0.4	0.8	0.8	114.2	0.4	177.0	1.1	11.6	15.4	2.6	1.6
4 threads	4.9	2.0	4.7	0.3	0.4	0.5	57.7	0.2	108.2	0.7	5.9	9.9	1.5	1.0

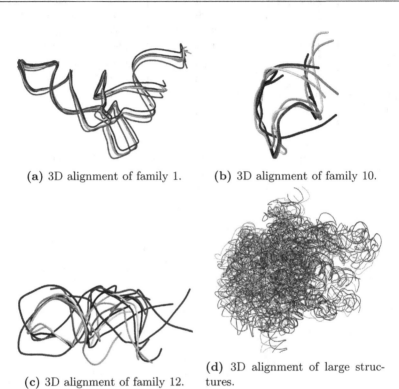

(a) 3D alignment of family 1.   **(b)** 3D alignment of family 10.

(c) 3D alignment of family 12.   **(d)** 3D alignment of large structures.

Fig. 4. 3D alignments

4.2 Runtime Evaluation

In this section we compare the time demands of MultiSETTER and SETTER classification. Reported times do not include time needed for PDB parsing, and for generating GSSUs since that can be done in advance in an initialization phase. To perform a classification by MultiSETTER, average structure of each family is created first. This is a one-time only task which is performed at the very beginning of the calculations. Times needed to produce average structures of individual families are presented in Tab. 2. The required time is proportional to the family size, ranging from less then one second for small families up to 177 seconds needed for family 7 with 17 structures. The average speedup when utilizing two cores (multithreaded) is about 75%. Finally, Tab. 3 shows times needed for the actual classification. We can observe that because of the precomputation of the families' average structures MultiSETTER is more than four times faster than SETTER. This speedup is even more significant when the parallel version of MultiSETTER is considered. For this purpose we utilized Intels Thread Building Blocks library. We parallelized the all-to-all distance matrix computation and also the GSSU merging part. When computing the all-to-all matrix, the individual distance computations are clearly independent. As for the merging, when the list of GSSU pairs is formed, then, again, the GSSU mergings of these pairs are independent and can thus be easily parallelized.

Table 3. Times in (s) needed to classify all the structures in the dataset and average time needed to classify one structure

Method	Time (s)	Avg. time (s)
SETTER	1669	16.7
MultiSETTER (1 thread)	359	3.6
MultiSETTER (4 threads)	160	1.6

5 Conclusion and Future Directions

In this paper we introduced a multiple structure alignment algorithm MultiSETTER built on top of the RNA structure pairwise alignment method SETTER. On the classification task utilizing structural data and annotations from the SCOR database we demonstrated that MultiSETTER yields better results than SETTER if SETTER is used to classify structures in a similar way as MultiSETTER, i.e. if it utilizes average structures of individual families. Due to the possibility to precompute average structures of individual families, MultiSETTER is also several fold faster than SETTER.

In the future we would like to focus on two aspects related to the RNA structure similarity domain. First is the improvement of the algorithm, mainly to improve its robustness with respect to the influence of the outliers. Second, we would like to apply MultiSETTER to automatically classify RNA structures. We would like to compare it to no longer supported SCOR manual classification, and to create an automated system of RNA structural classification.

Acknowledgments. This research was supported by the Czech Science Foundation (GAČR) project P202/11/0968 and by the Charles University project P46.

References

1. Eddy, S.R.: Non coding RNA genes and the modern RNA world. Nature Reviews Genetics 2(12), 919–929 (2001)
2. Bartel, D.P.: MicroRNAs: genomics, biogenesis, mechanism, and function. Cell 116, 281–297 (2004)

3. Hoksza, D., Svozil, D.: Efficient RNA pairwise structure comparison by setter method. Bioinformatics 28(14), 1858–1864 (2012)
4. Cech, P., Svozil, D., Hoksza, D.: Setter: web server for RNA structure comparison. Nucleic Acids Research 40(Web-Server-Issue), 42–48 (2012)
5. Kiryu, H., Tabei, Y., Kin, T., Asai, K.: Murlet: a practical multiple alignment tool for structural RNA sequences. Bioinformatics 23(13), 1588–1598 (2007)
6. Moretti, S., Wilm, A., Higgins, D.G., Xenarios, I., Notredame, C.: R-coffee: a web server for accurately aligning noncoding RNA sequences. Nucleic Acids Research 36(Web-Server-Issue), 10–13 (2008)
7. Katoh, K., Misawa, K., Kuma, K.I., Miyata, T.: MAFFT: a novel method for rapid multiple sequence alignment based on fast Fourier transform. Nucleic Acids Research 30(14), 3059–3066 (2002)
8. Torarinsson, E., Havgaard, J.H., Gorodkin, J.: Multiple structural alignment and clustering of RNA sequences. Bioinformatics 23(8), 926–932 (2007)
9. Tabei, Y., Kiryu, H., Kin, T., Asai, K.: A fast structural multiple alignment method for long RNA sequences. BMC Bioinformatics 9(1), 33 (2008)
10. Chang, Y.F., Huang, Y.L., Lu, C.L.: Sarsa: a web tool for structural alignment of RNA using a structural alphabet. Nucleic Acids Res. 36, 19–24 (2008)
11. Hendrix, D.K., Brenner, S.E., Holbrook, S.R.: RNA structural motifs: building blocks of a modular biomolecule. Quarterly Reviews of Biophysics 38(3), 221–243 (2005)
12. Holbrook, S.R.: Structural principles from large RNAs. Annual Review of Biophysics 37(1), 445–464 (2008)
13. Kabsch, W.: A solution for the best rotation to relate two sets of vectors. Acta Crystallographica Section A 32(5), 922–923 (1976)
14. Larkin, M.A., Blackshields, G., Brown, N.P., Chenna, R., McGettigan, P.A., Mcwilliam, H., Valentin, F., Wallace, I.M., Wilm, A., Lopez, R., Thompson, J.D., Gibson, T.J., Higgins, D.G.: ClustalW and ClustalX version 2.0. Bioinformatics 23, 2947–2948 (2007)
15. Saitou, N., Nei, M.: The neighbor-joining method: a new method for reconstructing phylogenetic trees. Molecular Biology and Evolution 4(4), 406–425 (1987)
16. Klosterman, P.S., Tamura, M., Holbrook, S.R., Brenner, S.E.: SCOR: a Structural Classification of RNA database. Nucleic Acids Res. 30(1), 392–394 (2002)
17. Tamura, M., Hendrix, D.K., Klosterman, P.S., Schimmelman, N.R., Brenner, S.E., Holbrook, S.R.: SCOR: Structural Classification of RNA, version 2.0. Nucleic Acids Res. 32(Database issue) (January 2004)

Assessing the Accuracy of the SIRAH Force Field to Model DNA at Coarse Grain Level

Pablo D. Dans[1,2], Leonardo Darré[1,3], Matías R. Machado[1], Ari Zeida[1,4],
Astrid F. Brandner[1], and Sergio Pantano[1,*]

[1] Institut Pasteur de Montevideo, Mataojo 2020, 11400, Uruguay
[2] Institute for Research in Biomedicine (IRB Barcelona), Baldiri Reixac, 10,
08028 Barcelona, Spain
[3] Department of Chemistry, King's College London, London, United Kingdom
[4] Deptartamento de Qca. Inorgánica, Analítica y Química-Física and INQUIMAE-CONICET,
Facultad de Ciencias Exactas y Naturales, Universidad de Buenos Aires, Ciudad Universitaria,
Pab. 2,C1428EHA Buenos Aires, Argentina
spantano@pasteur.edu.uy
www.sirahff.com

Abstract. We present a comparison between atomistic and coarse grain models
for DNA developed in our group, which we introduce here with the name
SIRAH. Molecular dynamics of DNA fragments performed using implicit and
explicit solvation approaches show good agreement in structural and dynamical
features with published state of the art atomistic simulations of double stranded
DNA (using Amber and Charmm force fields). The study of the multi-
microsecond timescale results in counterion condensation on DNA, in coincidence with high-resolution X-ray crystals. This result indicates that our model
for solvation is able to correctly reproduce ionic strength effects, which are very
difficult to capture by CG schemes.

Keywords: Molecular dynamics, nucleic acids, simulations, WT4, flexibility,
counterions, narrowing.

1   Introduction

Molecular Dynamics (MD) simulations have become a trustworthy and useful tool for
the study of the structural and dynamical behavior and interactions between biomolecules [1]. However, despite continuous developments [2-4], this technique is limited
to relatively small systems or short simulation times. Therefore, effort has been devoted to the implementation of simulation techniques based on the idea of simplified
or coarse grained (CG) representations of atomistic or fine grain (FG) systems, which
reduce significantly the computational demands but still capture the physical essence
of the phenomena under examination (see ref. [5] for a comprehensive review). Ow-
ing to its biological relevance, DNA has been subject of development of several CG

---

* Corresponding author.

J.C. Setubal and N.F. Almeida (Eds.): BSB 2013, LNBI 8213, pp. 71–81, 2013.

models (recently reviewed in [6]). Among others our group has developed a model for CG simulations of DNA [7] (Fig. 1a, b), which we present here for the first time with the acronym SIRAH (Southamerican Initiative for a Rapid and Accurate Hamiltonian). The SIRAH model can be used in combination with implicit or explicit solvation schemes using generalized Born model or a CG water model called WatFour [8] (WT4 for shortness, Fig. 1c). Moreover, it can be used for dual resolution simulations, in which FG and CG segments can be intercalated within the same DNA filament [9]. Here we present a critical assessment of the accuracy of our model using the Drew-Dickerson dodecamer (DD) [10] as main benchmark to compare our results with FG simulations and/or experimental data. The backmapping of CG coordinates allows recovering pseudo atomistic information from calculations performed at the CG level. Finally, we show that SIRAH can reproduce ionic strength effects and species-specific ionic binding to DNA, which are in agreement with high resolution X-ray data [11], and lead to a significant bending of the double helix.

2   Methods

2.1   The SIRAH Force Field for CG DNA and Aqueous Solvation

Our CG model for DNA (Fig. 1a, b) uses six effective beads per nucleobase, each placed in correspondence with the positions of real atoms in canonical conformations of FG nucleotides. A comparative view of the topology and excluded sizes of nucleotides, CG water and ions is shown in Fig. 1c. The partial charges on each bead add to a unitary negative charge on each nucleotide and ensure Watson-Crick electrostatic recognition. This charge distribution generates dipole moments, which are well compatible with those of state of the art FG force fields (Fig. 1d, e). A complete description of all parameters can be found in refs. [7,8]. In analogy with transient tetrahedral clusters formed by pure water, our model uses four beads interconnected in a tetrahedral conformation (Fig. 1c) [8]. Since each bead carries an explicit partial charge, WT4 liquid generates its own dielectric permittivity without the need to impose a uniform dielectric. The WT4 model reproduces several common properties of liquid water and simple electrolyte solutions [8].

2.2   MD Simulations and Analysis

The SIRAH model runs straightforwardly in the simulation packages AMBER and GROMACS (input/parameter files and tools to convert and visualize molecules are available from www.sirahff.com). Implicit solvent simulations, using the HCT pairwise generalized Born model [12] are performed with AMBER[13] using a cutoff of 18 Å and a salt concentration of 0.15 M. Temperature is controlled using a Langevin thermostat [14,15] with a friction constant of 50 ps^{-1}. Explicit solvent simulations are performed in the NPT ensemble using GROMACS 4.5[16]. A direct cut off for non-bonded interactions of 12 Å is used while long range electrostatics were evaluated using the PME approach [17]. Temperature and pressure are coupled to Berendsen thermostats and barostats [18] with coupling times of 1.2 ps and 6.0 ps, respectively. All systems are energy minimized and stabilized by raising the temperature from 0°K to 300°K in 1 ns.

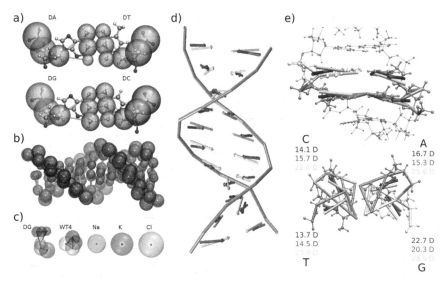

Fig. 1. SIRAH force field for CG DNA and aqueous solvent. a) FG nucleotides are presented with balls and sticks and colored by atoms (Oxygen: red, Nitrogen: blue, Carbon: cyan, Hydrogen: white). CG beads placed on FG positions are shown as semitransparent spheres. b) CG representation of the double helix DNA. c) Comparative view of CG molecules. Guanine, WT4 (water) and CG ions sodium, potassium and chloride are shown from left to right. Single electrolytes implicitly include a first solvation shell. The diameter of the CG beads in (a) and (c) corresponds to the actual van der Waals radii, providing an idea of the effective excluded volume and relative sizes of the beads. d) Schematic representation of dipole moments along the X-ray structure 1BNA showing the colinearity between FG and CG schemes (blue, red and yellow indicate Amber99-bsc0, Charmm27 and SIRAH force fields, respectively). e) Top: dipoles drawn on AT and GC base pairs showing their alignment with the base's planes. Bottom: same as top but seen along the DNA axis. Dipole modules are expressed in Debyes.

A time step of 20 fs is used. A list of the simulated systems and their composition are presented in Tab. 1. The X-ray structure of the Drew-Dickerson dodecamer (PDB id:1BNA [10]) was used as starting point. We compared our results against FG simulations on the same system performed with the Amber99-bsc0 [13] force field reported by Pérez et al [19] (sys3 in Tab. 1, available on-line at http://mmb.pcb.ub.es/microsecond/). Additionally, two systems bearing the 10 unique dinucleotide steps (namely, AA·TT, AC·GT, AG·CT, AT·AT, CA·TG, CC·GG, CG·CG, GA·TC, GC·GC and TA·TA) were simulated starting from the canonical B-form (sys5 and sys6 in Table 1) and compared with results reported in reference [20]. All the comparisons have been made on the back-mapped trajectories according to the procedure described in ref. [7]. Helical parameters are calculated using Curves+[21]. Root Mean Square Deviations (RMSD) are computed on all heavy atoms excluding the capping base pairs, while major and minor groove dimensions are measured between opposite phosphate groups and averaged along the double helix. Eigenvectors and eigenvalues are obtained by diagonalization of the covariance matrix calculated along the trajectories for all the heavy atoms using standard GROMACS utilities. As a gauge of the likeliness between different simulations, trajectories are fitted to

Table 1. Description of the simulated systems

System	Solvation model	n° solvent molecules [a]	Ionic Species (n° of ions)	Nucleotide sequence (5' to 3')	time (μs)
Sys1[c]	GB	---	---	CGCGAATTCGCG [b]	1.2
Sys1wr	GB	---	---	CGCGAATTCGCG [b]	1.2
Sys2[c]	WT4	523 (5753)	Na+(22)	CGCGAATTCGCG [b]	1.2
Sys2wr	WT4	523 (5753)	Na+(22)	CGCGAATTCGCG [b]	12
Sys3[d]	TIP3P	4998	Na+(22)	CGCGAATTCGCG [b]	1.2
Sys4	WT4	506 (5566)	Na+(19) K+(19) Cl-(16)	CGCGAATTCGCG [b]	12
Sys5	WT4	1510 (16610)	Na+(34) K+(33) Cl-(33)	GCCTATAAACGCCTATAA	10
Sys6	WT4	1510 (16610)	Na+(34) K+(33) Cl-(33)	CTAGGTGGATGACTCATT	10

[a] Parenthesis indicate the equivalent number of FG water molecules represented. [b] Drew-Dickerson dodecamer. [c] Simulated using harmonic constraints on the capping base-pairs. [d] Taken from ref.[19].

a common reference (the canonical structure) and their covariance matrices compared using a similarity index (SI) as in ref. [9].

The essential dynamics analysis is performed to compare sys1/sys2 with sys3. To avoid contaminating the main components of motion with possible helix fraying, in sys1 and sys2 constraints of 0.75 Kcal/mol•Å^2 are applied only to the Watson-Crick beads of capping bases. Counterions condensation is analyzed by computing electro-lyte occupancy density maps in 3D regular grids of 0.3 Å using VMD [22]. Cations closer than 5 Å to phosphate groups of both opposite strands are considered bound to the minor groove. Narrowing is measured only on the central track, i.e. between the four central phosphate pairs.

3   Results and Discussion

3.1   Structural and Dynamical Comparison

A first comparison of our CG model versus FG simulations (sys3) is performed in terms of RMSD on the backmapped trajectories of simulations using implicit and explicit solva-tion (sys1 and sys2, respectively). Along the 1,2 μs explored, both solvation schemes de-scribed equally well the DD structure with no RMSD drift from the experimental structure (Fig. 2a). In all the simulations the DNA duplexes show a flexible but stable behavior oscillating around the equilibrium B-form. The higher number of conformational substates explored by the FG simulation translates in higher RMSD variations. A good correspondence between the three simulations can be also inferred from structural super-position of backmapped snapshots taken at the beginning, middle and end of the dynamics (Fig. 2b).

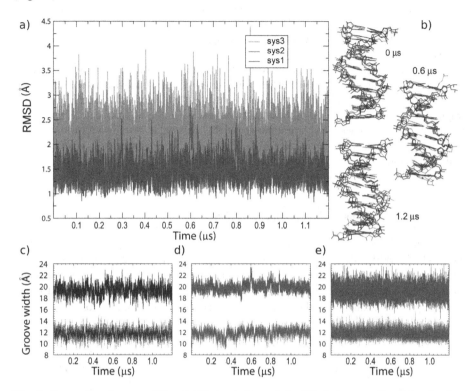

Fig. 2. Comparison between FG and CG simulations of the DD dodecamer. Implicit solvation CG (sys1), explicit solvation CG (sys2) and FG simulations (sys3) are presented in blue, red and green, respectively. a) RMSD along time calculated for all the heavy atoms (FG and back-mapped CG) respect to the X-ray structure 1BNA. b) Least mean square fit performed on all heavy atoms of conformers taken at the beginning, middle, and end of the trajectories. c - e) Major and minor groove widths (top and bottom traces, respectively) for the three systems.

Other characteristic features of DNA as the minor and major grooves show also a very good agreement with the FG simulation (Fig. 2c-e). In correspondence with the observation made for RMSD, both CG schemes show lower fluctuations. To gain a deeper insight on the dynamical behavior of the CG simulations, we compare the conformational sub-space sampled by inspecting the essential dynamics modes of each simulation. In all the cases, the first 3 eigenvectors explained nearly 50% of the total variance. These 3 essential modes are analyzed in more detail in terms of their projection onto the real space (see animation at http://www.youtube.com/watch?v=ivW7ixsG0fA&feature=youtu.be). The first mode involves a twisting and untwisting, the second is related with a simultaneous bending and twisting around the center of the AT track, while the third eigenvector, is associated to a global tilting of the duplex. To achieve a more quantitative characterization of the likeness between trajectories we calculate a similarity index (SI) from the covariance matrices of each simulation. As a measure for the maximum similarity reachable during the time window explored, we divided the FG trajectory in two halves and calculate the SI between both segments of trajectory resulting in a value of 0.91. Comparison of the FG versus the CG simulations results in SI values of 0.58 and 0.66 for implicit and explicit solvent, respectively. This roughly good similarity between the dynamics of FG and CG simulations may suggest that both approaches sample comparable potential energy landscapes.

To exclude the possibility that the loose constraints used on the capping base pairs may generate some artifacts, we perform analogous simulations without the restraints (sys1wr and sys2wr), which give equivalent results. The only difference is the helix fraying observed at both capping base pairs of sys1wr in the ns timescale. This behavior is in agreement with FG simulations for this particular system within the simulated time [19]. However, we have also reported on the spontaneous opening and rehybridization of longer DNA filaments in the multimicro second timescale, which produce no significant changes in the global structure of the double helix [23].

3.2   Base Pair Steps and Sequence Specificity

In previous publications we have reported sequence specific effects to influence melting temperatures and breathing profiles [7,23]. To achieve a more precise evaluation of the sequence-induced structural variations we simulated systems sys5 and sys6, which contain all the unique dinucleotide base pair steps. The helical parameters are compared upon backmapping with canonical values, averaged experiments and FG simulations (Fig. 3).

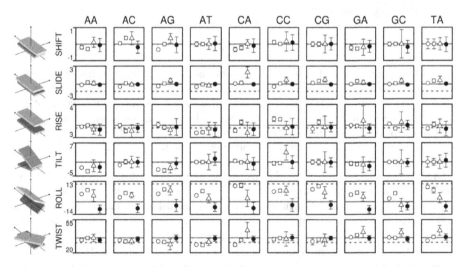

Fig. 3. Helical properties for the ten unique base pair steps. The helical properties SHIFT, SLIDE, RISE (measured in Å), TILT, ROLL and TWIST (measured in degrees) are compared for the force fields Amber99-bsc0 (empty circle), Charmm27 (squares), SIRAH (explicit solvation filled circles) and experimental x-ray measurements (triangles) including their standard deviations. All data except for that corresponding to SIRAH is taken from ref. [20]. Values of canonical B (blue line) and A (dashed red line) DNA forms are also given as references.

In general, the agreement with experiment and FG force fields is fairly good. Yet, the CG force field causes higher dispersion around average values when compared with FG simulations. The main deviation is observed for the ROLL, which shows a tendency to sample negative values. This problem is more marked for the steps CT and TC, which deviate almost 10 degrees from the canonical B-DNA. However, the impact of these deviations on the global structure and dynamics of the double helix seem to be minor, as judging from the results of the previous paragraphs.

3.3   Ionic Strength and DNA-Ion Binding

While the global distribution of cations around the DNA contributes to the stability of the double helix, the specific interaction of cations with DNA has been related with local structural distortions. In particular, the binding of sodium ions within the minor groove has been proposed to mediate its narrowing [24,25]. As quantified in our preceding publication [8], the binding of one single ion is enough to induce a sensible change in the minor groove. Increasing condensation of counterions translate in a progressively more marked minor groove narrowing.

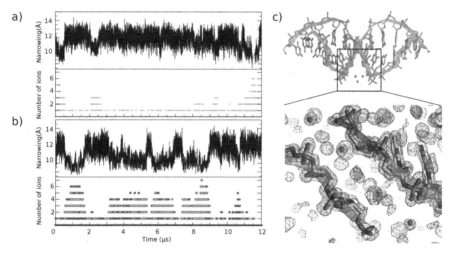

Fig. 4. Binding of cations within the minor groove of the DD dodecamer. a) Sys2wr simulation: minor groove width (top), and total number of bound cations in the minor groove as function of time. b) Idem to (a) for sys4. Green and blue dots are used for sodium and potassium, respectively. Red dots represent the sum of both electrolytes. c) Molecular representation of the DD dodecamer. The black square is zoomed in and rotated 90 degrees in the inset. Green and blue wireframes correspond to the occupational density calculated from the CG simulation for sodium and potassium ions, respectively. The gray wire frame shows the electron density from the X-ray data (PDB id: 355D [11]).

In the FG simulation (sys3) the simultaneous occupancy of the minor groove by several ions is very uncommon, but the presence of one Na^+ with residence times of 10 to 15 ns is not so rare [13]. The deformation observed in FG simulations on DNA by cations is insufficient to explain the distortions observed in the X-ray in presence of high salt concentration [11]. Two possible causes of this disagreement may be the lack of ionic strength (sys3 contains only neutralizing counterions) or insufficient sampling. To explore this issue we perform simulation sys2wr (containing only neutralizing counterions), increasing one order of magnitude the simulation time of sys2. Moreover, we also extended to 12 μs a previously published simulation [8] of the DD dodecamer in presence of added salts (sys4, Tab. 1). The simulation of sys2wr shows that binding of one single ion is very frequent and happens in the timescale from ns to μs (Fig. 4a). Conversely, we observe very few events where several sodium ions bind simultaneously into the minor groove. When these events happen, they have a longer duration and show a correlation with the narrowing of the minor groove. Simulation of the same DNA molecule in presence of added salts (sys4) offers a complementary view of this phenomenon.

The presence of added salts increases the degree of counterions condensation around the DNA (see animation at http://www.youtube.com/watch?v=kvIWQE8UcHo& feature=youtu.be). Both cations (Na+ and K+) localize around Phosphate moieties with particularly longer residence times into the minor groove (see also animation at http://www.youtube.com/watch?v=TapsTGisEew&feature=youtu.be). From a comparison between Figs. 4a and 4b it is clear that increasing the ionic strength in the solution changes sensibly the ionic binding profile presenting microseconds-long condensation events with the simultaneous binding of up to six counterions of both species present in the solution. Notably, the average narrowing of 9.6 Å measured during long condensation events, with 3 or more bound ions, coincides precisely with X-ray determinations [11,26]. Calculation of the occupational density of ions along the MD trajectory shows that the most populated occupational sites have a rough correspondence with the geometry reported for binding sites of water, sodium and potassium within the minor groove [8]. Superposition of the occupational density with crystallographic electron density shows that this agreement is particularly good for the atoms located between the phosphate moieties (Fig. 4c). This suggests that an extended counterion condensation is needed to generate a significant and sustained bending of the DNA [27]. Measuring the total bend with the program Curves+ [21] results in an average value of 26 degrees with extreme values ranging from 10 to 50 degrees. Using implicit solvent simulations with SIRAH, we recently reported that thermal oscillations lead to DNA breathing and formation of spontaneous kinks in the double helix [23]. The DNA bending angles of nearly 80% of the known universe of protein-DNA structures (curated in the PDI database [28], http://melolab.org/pdidb/web/content/links) falls within the values sampled by our simulations. Considering the present results, we notice that 62% of all the protein-DNA complexes present a bent minor or equal to the 26 degrees induced by ion binding found in this work. This leads to the intriguing conjecture that protein-DNA recognition might exploit spontaneous fluctuations driven by the electrolytic environment. Alternatively, one might think that the concomitant binding of ions creates low entropy regions in DNA, which are more prone to be targeted by protein ligands, considerably decreasing the free energy of binding.

4   Conclusions

We presented a systematic comparison of the performance of the SIRAH CG model for DNA in implicit and explicit solvation against FG simulations and experimental data. It turns out that both approximations provide a good description of the structural and dynamical features of DNA. The gross determinants of the structural stability of the double helix suggest that, upon backmapping, the information obtained from the CG simulations is almost as accurate as that provided by FG techniques. The specific and reversible binding of counterions within the minor groove generates microseconds-long narrowing, that probably relates to bent DNA conformations up to ~50 degrees. This distortion translated in a narrowing of the minor groove, which may be stable in the multi-microsecond time window.

The agreement with crystallographic data [26] and the comparison with all the high resolution protein-DNA complexes reported in the PDB provides a validation for our model and highlights the importance of the proper treatment of ionic strength effects. CG simulations sample a large range of DNA bending conformations seen in protein-DNA crystals, suggesting that, thermally induced oscillations of naked DNA encompasses the distortions required for protein binding. In line with similar conclusions conducted at the FG level [29], this add a new piece of evidence in favor of the prevalence of 'conformational selection' versus 'induced fit' paradigms.

The simulation schemes presented here result in a speed up respect to their corresponding FG simulations of 800 and 2400 times for explicit and implicit solvent simulations, respectively. Since the high computational cost associated to atomistic simulations precludes the study of many interesting phenomena, this kind of approaches is expected to become of regularly use and interest for the broad scientific community.

Acknowledgements. This work was supported by ANII – Agencia Nacional de Investigación e Innovación, Programa de Apoyo Sectorial a la Estrategia Nacional de Innovación – INNOVA URUGUAY (Agreement n8 DCI – ALA / 2007 / 19.040 between Uruguay and the European Commission). A.B. is beneficiary of the National Fellowship System of ANII. P.D.D. and S.P. appreciate support from the National Scientific Program of ANII (SNI) and PEDECIBA. We thank Adrian Roitberg for useful discussions about the implementation of the WT4 topology in the AMBER package and Sebastian Ferreira for excellent technical support.

References

[1] Karplus, M., McCammon, J.A.: Molecular dynamics simulations of biomolecules. Nat. Struct. Biol. 9, 646–652 (2002)

[2] Shaw, D.E., Maragakis, P., Lindorff-Larsen, K., Piana, S., Dror, R.O., Eastwood, M.P., Bank, J.A., Jumper, J.M., Salmon, J.K., Shan, Y., Wriggers, W.: Atomic-level characterization of the structural dynamics of proteins. Science 330, 341–346 (2010)

[3] Arkhipov, A., Yin, Y., Schulten, K.: Four-scale description of membrane sculpting by BAR domains. Biophys. J. 95, 2806–2821 (2008)

[4] Yin, Y., Arkhipov, A., Schulten, K.: Simulations of membrane tubulation by lattices of amphiphysin N-BAR domains. Structure 17, 882–892 (2009)

[5] Voth, G.A.: Coarse-Graining of Condensed Phase and Biomolecular Systems. Taylor & Francis Group, New-York (2009)

[6] Potoyan, D., Savelyev, A., Papoian, G.: Recent successes in coarse-grained modeling of DNA. WIREs Comput. Mol. Sci. 3, 69–83 (2013)

[7] Dans, P.D., Zeida, A., Machado, M.R., Pantano, S.: A Coarse Grained Model for Atomic-Detailed DNA Simulations with Explicit Electrostatics. J. Chem. Theory Comput. 6, 1711–1725 (2010)

[8] Darré, L., Machado, M.R., Dans, P.D., Herrera, F.E., Pantano, S.: Another Coarse Grain Model for Aqueous Solvation: WAT FOUR? J. Chem. Theory Comput. 6, 3793–3807 (2010)

[9] Machado, M.R., Dans, P.D., Pantano, S.: A hybrid all-atom/coarse grain model for multiscale simulations of DNA. Phys. Chem. Chem. Phys. 13, 18134–18144 (2011)

[10] Drew, H.R., Wing, R.M., Takano, T., Broka, C., Tanaka, S., Itakura, K., Dickerson, R.E.: Structure of a B-DNA dodecamer: conformation and dynamics. Proc. Natl. Acad. Sci. U. S. A. 78, 2179–2183 (1981)

[11] Shui, X., McFail-Isom, L., Hu, G.G., Williams, L.D.: The B-DNA dodecamer at high resolution reveals a spine of water on sodium. Biochemistry 37, 8341–8355 (1998)

[12] Hawkins, G.D., Cramer, C.J., Truhlar, D.G.: Parametrized models of aqueous free energies of solvation based on pairwise descreening of solute atomic charges from a dielectric medium. J. Phys. Chem. 100, 19839 (1996)

[13] Perez, A., Marchan, I., Svozil, D., Sponer, J., Cheatham III, T.E., Laughton, C.A., Orozco, M.: Refinement of the AMBER force field for nucleic acids: improving the description of alpha/gamma conformers. Biophys. J. 92, 3817–3829 (2007)

[14] Pastor, R.W., Brooks, B.R., Szabo, A.: An analysis of the accuracy of Langevin and molecular dynamics algorithms. Mol. Phys. 65, 1409–1419 (1988)

[15] Wu, X., Brooks, B.R.: Self-guided Langevin dynamics simulation method. Chem. Phys. Lett. 381, 512–518 (2003)

[16] Hess, B., Kutzner, C., van de Spoel, D., Lindahl, E.: GROMACS 4: Algorithms for Highly Efficient, Load-Balanced, and Scalable Molecular Simulation. J. Chem. Theo. Comp. 4, 435–447 (2008)

[17] Essmann, U., Perera, L., Berkowitz, M.L., Darden, T.A., Lee, H., Pedersen, L.: A smooth particle mesh ewald potential. J. Chem. Phys. 103, 8577–8592 (1995)

[18] Berendsen, H.J.C., Postma, J.P.M., van Gunsteren, W.F., DiNola, A., Haak, J.R.: Molecular dynamics with coupling to an external bath. J. Chem. Phys. 81, 3684–3691 (1984)

[19] Pérez, A., Luque, F.J., Orozco, M.: Dynamics of B-DNA on the Microsecond Time Scale. J. Am. Chem. Soc. 129, 14739–14745 (2007)

[20] Perez, A., Lankas, F., Luque, F.J., Orozco, M.: Towards a molecular dynamics consensus view of B-DNA flexibility. Nucleic Acids Res. 36, 2379–2394 (2008)

[21] Lavery, R., Moakher, M., Maddocks, J.H., Petkeviciute, D., Zakrzewska, K.: Conformational analysis of nucleic acids revisited: Curves+. Nucleic Acids Res. 37, 5917–5929 (2009)

[22] Humphrey, W., Dalke, A., Schulten, K.: VMD - Visual Molecular Dynamics. J. Molec. Graphics 14, 33–38 (1996)

[23] Zeida, A., Machado, M.R., Dans, P.D., Pantano, S.: Breathing, bubbling, and bending: DNA flexibility from multimicrosecond simulations. Phys. Rev. E Stat. Nonlin. Soft. Matter Phys. 86, 021903 (2012)

[24] Hamelberg, D., Williams, L.D., Wilson, W.D.: Influence of the dynamic positions of cations on the structure of the DNA minor groove: sequence-dependent effects. J. Am. Chem. Soc. 123, 7745–7755 (2001)

[25] Hamelberg, D., Williams, L.D., Wilson, W.D.: Effect of a neutralized phosphate backbone on the minor groove of B-DNA: molecular dynamics simulation studies. Nucleic Acids Res. 30, 3615–3623 (2002)

[26] Shui, X., Sines, C.C., McFail-Isom, L., VanDerveer, D., Williams, L.D.: Structure of the potassium form of CGCGAATTCGCG: DNA deformation by electrostatic collapse around inorganic cations. Biochemistry 37, 16877–16887 (1998)

[27] Spiriti, J., Kamberaj, H., de Graff, A., Thorpe, M.F., van der Vaart, A.: DNA Bending through Large Angles Is Aided by Ionic Screening. J. Chem. Theo. Comp. 8, 2145–2156 (2012)

[28] Norambuena, T., Melo, F.: The Protein-DNA Interface database. BMC Bioinformatics 11, 262 (2010)

[29] Dans, P.D., Perez, A., Faustino, I., Lavery, R., Orozco, M.: Exploring polymorphisms in B-DNA helical conformations. Nucleic Acids Res. 40, 10668–10678 (2012)

How to Multiply Dynamic Programming Algorithms

Christian Höner zu Siederdissen[1], Ivo L. Hofacker[1−3],
and Peter F. Stadler[4,1,3,5−7]

[1] Dept. Theoretical Chemistry, Univ. Vienna, Währingerstr. 17, Wien, Austria
[2] Bioinformatics and Computational Biology Research Group, University of Vienna,
1090 Währingerstraße 17, Vienna, Austria
[3] RTH, Univ. Copenhagen, Grønnegårdsvej 3, Frederiksberg C, Denmark
[4] Dept. Computer Science, and Interdisciplinary Center for Bioinformatics,
Univ. Leipzig, Härtelstr. 16-18, Leipzig, Germany
[5] MPI Mathematics in the Sciences, Inselstr. 22, Leipzig, Germany
[6] FHI Cell Therapy and Immunology, Perlickstr. 1, Leipzig, Germany
[7] Santa Fe Institute, 1399 Hyde Park Rd., Santa Fe, USAx

Abstract. We develop a theory of algebraic operations over linear grammars that makes it possible to combine simple "atomic" grammars operating on single sequences into complex, multi-dimensional grammars. We demonstrate the utility of this framework by constructing the search spaces of complex alignment problems on multiple input sequences explicitly as algebraic expressions of very simple 1-dimensional grammars. The compiler accompanying our theory makes it easy to experiment with the combination of multiple grammars and different operations. Composite grammars can be written out in LaTeX for documentation and as a guide to implementation of dynamic programming algorithms. An embedding in Haskell as a domain-specific language makes the theory directly accessible to writing and using grammar products without the detour of an external compiler.
http://www.bioinf.uni-leipzig.de/Software/gramprod/

Keywords: linear grammar, context free grammar, product structure, multiple alignment, Haskell.

1 Introduction

The well-known dynamic programming algorithms for the simultaneous alignment of n sequences [1] have a structure that is reminiscent of topological product structures. This is expressed e.g. by the fact that intermediary tables are n-dimensional. Here we explore if and how this intuition can be made precise and operational. To this end we build on the conceptual framework of Algebraic Dynamic Programming (ADP) [2, 3]. In this setting a dynamic programming (DP) algorithm is separated into a context-free grammar (CFG) that generates the search space and an evaluation algebra. In this contribution we will mainly be concerned with a notion of product grammars to facilitate the construction of the search space.

J.C. Setubal and N.F. Almeida (Eds.): BSB 2013, LNBI 8213, pp. 82–93, 2013.

Before we delve into a more formal presentation, consider the context-free grammar for pairwise sequence alignment with affine gap costs as an example. Gotoh's algorithm [4] uses three non-terminals M, D, I, depending on whether the right end of the alignment is a match state, a gap in the first sequence, or a gap in the second sequence. The corresponding productions are of the form

$$
\begin{aligned}
M &\to M(\tbinom{u}{v}) \mid D(\tbinom{u}{v}) \mid I(\tbinom{u}{v}) \mid (\tbinom{\varepsilon}{\varepsilon}) \\
D &\to M(\tbinom{u}{\_}) \mid D(\tbinom{u}{\cdot}) \mid I(\tbinom{u}{\_}) \\
I &\to M(\tbinom{-}{v}) \mid D(\tbinom{-}{v}) \mid I(\tbinom{\cdot}{v})
\end{aligned}
\tag{1}
$$

where u and v denote terminal symbols, '$-$' corresponds to gap opening, while '.' denotes the (differently scored) gap extension. The ε here takes the role of the "sentinel character", i.e., matches the end of the input. Each of the non-terminals reads simultaneously from two separate input tapes. To make this property more transparent in the notation, we write $M \rightsquigarrow (\tbinom{X}{X})$, $D \rightsquigarrow (\tbinom{X}{Y})$, and $I \rightsquigarrow (\tbinom{Y}{X})$. This yields productions such as

$$
(\tbinom{X}{X}) \to (\tbinom{X}{X})(\tbinom{u}{v}) \simeq (\tbinom{Xu}{Xv}) \qquad \text{or} \qquad (\tbinom{Y}{X}) \to (\tbinom{X}{Y})(\tbinom{-}{v}) \simeq (\tbinom{X-}{Yv})
\tag{2}
$$

Apart from the conspicous absence of $(\tbinom{Y}{Y})$, i.e., alignments ending in an all-gap column, to which we will return later, this notation strongly suggests to consider the 1-dimensional projections of the 2-dimensional productions of Equ. (2), which obviously have the form

$$
X \to Xu \mid Yu \mid \varepsilon \qquad \text{and} \qquad Y \to Y. \mid X-
\tag{3}
$$

This simple grammar either reads a symbol (non-terminal X) or it ignores it (non-terminal Y). Each copy of the "step grammar" (3) operates on its own input tape. The basic idea in this contribution is to consider the dynamic programming algorithms for n-way alignments as an n-fold product of the simple step grammar with itself. To this end we need to solve two problems: First, we need to clarify the precise meaning of the product of CFGs. Since alignment algorithms are naturally expressed as left-linear CFGs we will be content with this special case here. Second, we need to develop a theory for the construction of the evaluation algebra for a product grammar.

We note that full-fledged n-way DP alignments have exponential running time and hence are of little practical use for large n. Although elaborate divide & conquer strategies have been proposed to prune the search space, see e.g. [1], heuristic approaches that combine pairwise alignments are much more common. Three-way alignments nevertheless are employed in practise in particular when high accuracy is crucial, see e.g. [5–8]. Four-way alignments were recently explored for aligning short words from human language data [9].

2 Algebraic Operations on Grammars

2.1 Notation

A CFG $\mathcal{G} = (N, T, P, s)$ consists of a finite set N of non-terminals, a finite set T of terminals so that $N \cap T = \emptyset$, a set P of productions $x \to \alpha$ where $x \in N$

and $\alpha \in (T \cup N)^*$, and a start symbol $s \in N$. Furthermore, we need the special symbol ϵ denoting the empty string and an "empty production" $\varnothing$. Throughout the body of this contribution we will consider in particular left-linear grammars, i.e., those for which all productions are of the form $A \to Bx$ with $A, B \in N$ and $x \in T$.

The example of Gotoh's algorithm in the introductory section motivates us to introduce algebraic operations on grammars in a more systematic way. As a running example, we will use one of the simplest alignment algorithms. The Needleman-Wunsch algorithm [10] aligns two sequences $x_{1...n}$ and $y_{1...m}$ so that the sum of match and in/del scores is maximized. The basic recursion over the memoization table T reads

$$T_{ij} = \max \left\{ T_{i-1,j} + d, T_{i,j-1} + d, T_{i-1,j-1} + m(x_i, y_j), 0_{i=0,j=0}, -\infty \right\} \quad (4)$$

In the recursive scheme, the base case is given by the alignment of two empty substrings "on the left", while the other cases extend the already aligned part of the strings to the right. The first two cases denote an in/del operation with cost d, while $m(.,.)$ scores the (mis)match x_i with y_j.

A two-tape grammar equivalent to the recursion in Equ. (4) is

$$\left(\begin{smallmatrix} X \\ Y \end{smallmatrix} \right) \to \left(\begin{smallmatrix} X \\ Y \end{smallmatrix} \right)\left(\begin{smallmatrix} a \\ \epsilon \end{smallmatrix} \right) \mid \left(\begin{smallmatrix} X \\ Y \end{smallmatrix} \right)\left(\begin{smallmatrix} \epsilon \\ a \end{smallmatrix} \right) \mid \left(\begin{smallmatrix} X \\ Y \end{smallmatrix} \right)\left(\begin{smallmatrix} a \\ a \end{smallmatrix} \right) \mid \left(\begin{smallmatrix} \epsilon \\ \epsilon \end{smallmatrix} \right) \quad (5)$$

There are several differences between the formulation in Equ. (4) and Equ. (5). The recursive formulation working on the memoization table T does not store the alignment directly but rather the *score* of each partial, optimal alignment. The grammatical description, on the other hand, describes the *search space* of all possible alignments without any notation of scoring. In addition, recursive descriptions usually include explicit annotations for base cases, here the empty alignment. The production rule $\left(\begin{smallmatrix} X \\ Y \end{smallmatrix} \right) \to \left(\begin{smallmatrix} \epsilon \\ \epsilon \end{smallmatrix} \right)$ has this role in our example. In general, grammatical descriptions abstract away certain implementation details. Some of these will, however, become important when constructing more complex grammars from simpler ones, as we shall see below.

Our task will be to construct the Equ. (5) from even simpler, "atomic" constituents. These grammars are

$$\mathcal{S} = (\{X\}, \{a\}, \{X \to Xa \mid X\}, X) \quad (6)$$
$$\mathcal{N} = (\{X\}, \{\varepsilon\}, \{X \to \varepsilon\}, X) \quad (7)$$
$$\mathcal{L} = (\{X\}, \{\}, \{X \to X\}, X) \quad (8)$$

The grammar $\mathcal{S}$ in Equ. (6) performs a "step". It either reads a single character on the right and recurses on the left, or simply recurses. Note that by itself the rules do not terminate. The grammar $\mathcal{N}$, Equ. (7), matches the empty input (or any empty substring of the input) and immediately terminates. Finally, $\mathcal{L}$ Equ. (8) reproduces the non-terminating loop case already seen in Equ. (6).

Intuitively, we can combine these three components on a single tape as

$$\mathcal{S} + \mathcal{N} - \mathcal{L} = (\{X\}, \{a, \varepsilon\}, \{X \to Xa \mid \varepsilon\}, X) \quad (9)$$

In order to make this intuition precise we need to give a precise meaning to algebraic operations on grammars. In the following we will do this for linear grammars, however with an extension to general CFGs in mind.

Each operator introduced below primarily acts on sets of production rules. They implicitly carry over to the involved sets of terminals and non-terminals in an obvious manner. Two production rules are equivalent if they are isomorphic as in Equ. (13). This is of relevance insofar that it leads to idempotency in one of the operators below, but does not otherwise interfere with parsing[1]. In the following we use the notation P^n to emphasize that the productions operate on n tapes. We will refer to $\dim \mathcal{G} = n$ as the dimension of the grammar.

2.2   Algebraic Operations on Grammars

The + monoid. The + operator is defined as the union of all production rules of the two grammars:

$$P_1^n + P_2^n = P_1^n \cup P_2^n \tag{10}$$

We enforce explicitly that the + operator requires that the two operand grammars have the same dimensionality. The + operation forms a monoid over the set of production rules. Since the production rules form a set, isomorphic rules collapse to a single rule. The empty set $P^n = \{\}$ is a neutral element and $P^n + P^n = P^n$, i.e., the + monoid is idempotent. Isomorphism on production rules is also symbolic, that is, $X \to X$ is isomorphic to $X \to X$ but not to $\{X \to Y, Y \to X\}$, even though the latter set of two rules reduces to the first. For our example, we have $(X \to Xa \mid X) + (X \to \epsilon) = (X \to Xa \mid X \mid \epsilon)$.

The − operator. While the + operator unifies two sets of production rules, the − operator acts as a set difference operator

$$P_1^n - P_2^n = \{p \in P_1^n | p \notin P_2^n\} \tag{11}$$

As for +, it requires operands of the same dimensionality. By construction, − is not associative. Thus does not form a semigroup but merely a magma. The empty set of production rules acts as the neutral element on the right. This operator is important to explictly remove production rules that yield infinite derivations. In our example, we need to remove $\{X \to X\}$. With the help of − we can write $(X \to Xa \mid X) - (X \to X) = (X \to Xa)$. We shall see that it is often convenient to "temporarily" introduce productions that later on need to be excluded from the final algorithm.

The ⊗ monoid. The definition of a direct product of left linear grammars lies at the heart of this contribution.

Definition 1. *Let $\mathcal{G}_1 = (N_1, T_1, P_1, s_1)$ and $\mathcal{G}_2 = (N_2, T_2, P_2, s_2)$ be left-linear CFGs, i.e., all productions are of the form $A \to Bx$ or $A \to y$. Their direct*

---

[1] This is not completely true in the context of stochastic linear grammars: replication of a rule in an SCFG that already has duplicated rules requires that we sum over the probabilities for isomorphic rules.

product $\mathcal{G}_1 \otimes \mathcal{G}_2$ is the grammar $\mathcal{G} = (N, T, P, s)$ with non-terminals $N = N_1 \times N_2 \cup N_1 \times \{\epsilon\} \cup \{\epsilon\} \times N_2$, terminals $T = T_1 \times T_2 \cup T_1 \times \{\epsilon\} \cup \{\epsilon\} \times T_2$, the start symbol of the product is $s = \binom{s_1}{s_2}$. The productions are of the forms $\binom{A_1}{A_2} \to \binom{B_1}{B_2}\binom{x_1}{x_2}$, $\binom{A_1}{A_2} \to \binom{B_1}{\epsilon}\binom{x_1}{y_2}$, $\binom{A_1}{A_2} \to \binom{\epsilon}{B_2}\binom{y_1}{x_2}$, $\binom{A_1}{A_2} \to \binom{y_1}{y_2}$, $\binom{A_1}{\epsilon} \to \binom{B_1}{\epsilon}\binom{x_1}{\epsilon}$, $\binom{\epsilon}{A_2} \to \binom{\epsilon}{B_2}\binom{\epsilon}{x_2}$, $\binom{A_1}{\epsilon} \to \binom{y_1}{\epsilon}$, and $\binom{\epsilon}{A_2} \to \binom{\epsilon}{y_2}$ iff $A_1 \to B_1 x_1$ and $A_1 \to y_1$, are productions in P_1 and $A_2 \to B_2 x_2$, $A_2 \to y_2$ are productions in P_2, respectively.

By construction $\mathcal{G}$ is again a left-linear CFG that now operates on two bands. It will be convenient to abuse the notation and write productions of the form $A_i \to y_i$ as $A_i \to \epsilon y_i$. Hence all productions in the product grammar can be written as $\binom{A_1}{A_2} \to \binom{B_1}{B_2}\binom{x_1}{x_2}$ with $A_i, B_i \in N_i \cup \{\epsilon\}$, $x_i \in T_i \cup \{\epsilon\}$ subject to the following conditions: $A_i = \epsilon$ implies $B_i = x_i = \epsilon$, $\binom{A_1}{A_2} \neq \binom{\epsilon}{\epsilon}$, and $\binom{\epsilon}{\epsilon}$ on the r.h.s. is omitted. We will also make use of notation $(A_1 \to B_1 y_1) \otimes (A_2 \to B_2 y_2)$ for the product of two individual productions. By construction, we have

$$\dim(\mathcal{G}_1 \otimes \mathcal{G}_2) = \dim \mathcal{G}_1 + \dim \mathcal{G}_2 \tag{12}$$

We note finally, that the empty string ϵ appearing in the 2-dimensional terminals and non-terminals is not necessarily associated with terminating the reading from the input band(s).

To see that $\otimes$ is associative we need to demonstrate that the productions on $(\mathcal{G}_1 \otimes \mathcal{G}_2) \otimes \mathcal{G}_3$ and $\mathcal{G}_1 \otimes (\mathcal{G}_2 \otimes \mathcal{G}_3)$ are isomorphic, i.e.,

$$\left(\binom{x_1}{x_2}{x_3} \right) \to \left(\binom{\alpha_1}{\alpha_2}{\alpha_3} \right) \quad \simeq \quad \binom{x_1}{x_2}{x_3} \to \left(\binom{\alpha_1}{\alpha_2}{\alpha_3} \right) \tag{13}$$

This is most easily seen in the notation with the extra ϵ since in this case the α_i are strings of length 2 that are simply decomposed columnwise. Hence multiple products are well-defined. Furthermore, permutations of rows are isomorphisms. Thus $\mathcal{G}_1 \otimes \mathcal{G}_2 \simeq \mathcal{G}_2 \otimes \mathcal{G}_1$, i.e. exchanging the order of factors affects the order of the coordinates only. Due to the associativity of $\otimes$, we can safely extend these constructions to more than two factors.

The canonical projection $\pi_i : \mathcal{G}_1 \otimes \mathcal{G}_2 \to \mathcal{G}_i$ is obtained by formally isolating the i-th coordinate and contracting the empty strings ϵ and the empty productions $\varnothing = (\epsilon \to \epsilon)$. Clearly we have $\pi_i(T) = T_i$, $\pi_i(N) = N_i$, $\pi_i(s) = s_i$, and $\pi_i(P) = P_i$. The grammar product $\otimes$ thus has the basic properties of a well-defined product.

Let $\text{lan}(\mathcal{G})$ denote the language generated by G. Note that a "string" in $\text{lan}(\mathcal{G})$ is, by construction, a sequence of terminals, each of which is either of the form $\binom{x_1}{x_2}$ with $x_1 \in T_1$ and $x_2 \in T_2$, or of the form $\binom{x_1}{\epsilon}$ with $x_1 \in T_1$, or of the form $\binom{\epsilon}{x_2}$. Thus $\text{lan}(\mathcal{G}_1 \otimes \mathcal{G}_2)$ consists of alignments of strings $\alpha_i \in \mathcal{G}_i$. To see this, note that each string $\alpha_i \in \mathcal{G}_i$ is generated from s_i using a finite sequence $\wp_i = (p_i^1, p_i^2, \dots)$ of productions. Any partial matching of the $\wp_1$ and $\wp_2$ that preserves the sequential order of the two input sequences gives rise to a sequence of productions $\wp \in P^*$ by matching all unmatched p_i^k with the dummy production $\varnothing$. By construction $\pi_i(\wp) = \wp_i$, i.e., $\wp$ derived an alignment of the input strings β_1 and β_2. Conversely, given a sequence $\wp$ of productions of the product grammar, we know that $\pi_i(\wp)$ is a sequence of productions of $\mathcal{G}_i$; hence it constructs strings in $\text{lan}(\mathcal{G}_i)$. It follows that the product language satisfies

$$\pi_i(\text{lan}(\mathcal{G}_1 \otimes \mathcal{G}_2)) = \text{lan}(\mathcal{G}_i) \tag{14}$$

Similarly, we find that parse trees have a natural alignment structure. Let τ be a parse tree for an input $\beta \in \text{lan}(\mathcal{G}_1 \times \mathcal{G}_2)$. Its interior nodes are labeled by the productions, i.e., pairs of the form $\begin{pmatrix} A_1 \rightarrow B_1 x_1 \\ A_2 \rightarrow B_2 x_2 \end{pmatrix}$, $\begin{pmatrix} A_1 \rightarrow B_1 x_1 \\ \varnothing \end{pmatrix}$, or $\begin{pmatrix} \varnothing \\ A_2 \rightarrow B_2 x_2 \end{pmatrix}$. The projections $\pi_i(\tau)$ are explained by retaining only the i-th coordinate of the vertex label and contracting all vertices labeled by $\varnothing$ in $\pi_i(\tau)$ yields a valid parse tree for $\pi_i(\beta)$ w.r.t. $\mathcal{G}_i$. Thus τ is a tree alignment of the parse trees for the two input strings.

The direct product $\otimes$ forms a monoid on grammars with arbitrary dimensions since

$$P_1^m \otimes P_2^n = \{(p_1 \otimes p_2)^{m+n} | p_1^m \in P_1^m, p_2^n \in P_2^n\}, \tag{15}$$

where $p_1 \otimes p_2$ is explained in Def. 1. The neutral element of the $\otimes$ monoid is the zero-dimensional grammar which has one production rule $\epsilon^0 \rightarrow \epsilon^0$ that neither reads nor writes anything as it does not operate on a tape. Albeit rather artificial at first glance, it is useful to have a neutral element available. For our example, we have

$$(X \rightarrow Xa|X) \otimes (X \rightarrow Xa|X)$$
$$= \begin{pmatrix} X \\ X \end{pmatrix} \rightarrow \begin{pmatrix} X \\ X \end{pmatrix}\begin{pmatrix} a \\ a \end{pmatrix} \mid \begin{pmatrix} X \\ X \end{pmatrix}\begin{pmatrix} a \\ \epsilon \end{pmatrix} \mid \begin{pmatrix} X \\ X \end{pmatrix}\begin{pmatrix} \epsilon \\ a \end{pmatrix} \mid \begin{pmatrix} X \\ X \end{pmatrix} \tag{16}$$

This grammar contains the 2-dimensional loop rule $\begin{pmatrix} X \\ X \end{pmatrix} \rightarrow \begin{pmatrix} X \\ X \end{pmatrix}$, derived from $(X \rightarrow X) \otimes (X \rightarrow X)$ that eventually needs to be eliminated. To this end, it will be convenient to consider yet another operation on productions.

The structure-preserving power $*$ For any k-dimensional grammar $\mathcal{G}$ and any natural number $n \in \mathbb{Z}$, $\mathcal{G} * n$ denotes the $k \times n$-dimensional grammar with the same structure. Each k-dimensional (terminal or non-terminal) symbol $\begin{pmatrix} s_1 \\ s_k \end{pmatrix}$ is transformed to an $k \times n$-dimensional symbol $\begin{pmatrix} \begin{pmatrix} s_1 \\ s_k \end{pmatrix} \\ \begin{pmatrix} s_1 \\ s_k \end{pmatrix} \end{pmatrix}$. Note that for a grammar with a single production rule we have $G \otimes G \equiv G * 2$.

For our example grammar, this operation is useful as short-hand for both Equ. 7 and Equ. 8. In the case of linear grammars, the $*$ operator is mostly useful as shorthand to expand singleton grammars. It is worth noting, however, that a number of algorithms, notably [11], in computational biology work on multiple tapes with a grammar structure equal to their one-dimensional cousins. In particular, the Sankoff algorithm [11] is a variant of the Nussinov algorithm extended to two tapes.

We can now construct the full Needleman-Wunsch alignment grammar from the much simpler 1-dimensional constituents of Eqns.(6–8) in the following way:

$$\mathcal{NW} = \mathcal{G} \otimes \mathcal{G} + \mathcal{N} * 2 - \mathcal{L} * 2, \tag{17}$$

Written in terms of the productions only, this can be rephrased as

$$(X \rightarrow Xa|X) \otimes (X \rightarrow Xa|X) + (X \rightarrow \epsilon) * 2 - (X \rightarrow X) * 2$$
$$= \begin{pmatrix} X \\ X \end{pmatrix} \rightarrow \begin{pmatrix} X \\ X \end{pmatrix}\begin{pmatrix} a \\ a \end{pmatrix} \mid \begin{pmatrix} X \\ X \end{pmatrix}\begin{pmatrix} a \\ \epsilon \end{pmatrix} \mid \begin{pmatrix} X \\ X \end{pmatrix}\begin{pmatrix} \epsilon \\ a \end{pmatrix} \mid \begin{pmatrix} \varepsilon \\ \varepsilon \end{pmatrix} \tag{18}$$

Note that we have used here a distinct symbol ε to highlight the termination case deriving from $\mathcal{N}$. Since our construction of the Needleman-Wunsch grammar is based on well-defined algebraic operations we can readily use the same approach to construct much more complex alignment algorithms. Before we proceed, however, we need to address the technical issue of loop rules.

2.3   Grammars with Loops

In Equ. (17) we explicitly added a terminating base case $X \to \varepsilon$ and removed a production rule with infinite derivations $X \to X$. Why do we insist on performing this operation explicitly instead of modifying the definition of the direct product $\otimes$ accordingly?

The main reason lies in performance considerations. An "intelligent" product operator would first need to determine which rules have infinite derivations. For linear grammars with only one non-terminal a rule is not infinite if a single terminal (except ϵ) is present. ϵ rules are also fine, as long as only the empty word case $X \to \epsilon$ is present. Productions of the form $\{X \to Y, Y \to X\}$, however need to be followed up to a depth of the number of production rules present. For context-free grammars, the complexity will increase further, as now multiple non-terminals may exist on the right-hand side. For both convenience and efficiency (by a constant factor), it does not seem to be desirable to transform the grammar into Chomsky normal form. The second problem is the need for rewriting. In the case of $\{X \to Y, Y \to X\}$, rewriting yields $X \to X$ by inserting the rules for Y wherever Y is used. More complicated grammars might quite easily require major rewrites before all loop cases can be removed.

Finally, using looping productions can be conceptually useful during construction. In case of Equ. 6, we either want to read a character in a "step" $X \to Xa$ or perform an in/del with a "stand" $X \to X$. The direct product of Equ. (6) then yields all possibilities of stepping or standing on two (or more) tapes. Of these cases we only want to remove the case where all tapes "stand". This case is quite easily determined as Equ. 8 and just needs to be scaled (with $*$) to the correct dimension and subtracted from the complete grammar.

2.4   Implementation

We have implemented a small compiler for our grammar product formalism with three output targets. First, we generate LaTeX output. This supports researchers in the development of complex, multiple dimensional linear grammars, facilitates the comparison with the intended model for an elaborate alignment-like algorithm. It assists implementation of the grammar in the users' programming language of choice as the mathematical description of the recurrences reduces the chance that a production rule or recursion is simply forgotten.

In addition, we directly target the functional programming language Haskell [12]. It is possible to emit a Haskell module prototype which then needs to be extended with user-defined evaluation (scoring) algebras. This mode mirrors the LaTeX output. Advanced users may make use of TemplateHaskell [13] to *directly embed* our domain-specific language as a proper extension of Haskell itself. Both Haskell-based approaches ultimately make use of stream fusion optimizations

[14] by way of the `ADPfusion` [15] framework that produces efficient code for dynamic programming algorithms.

Currently, the emitted Haskell code for non-trivial applications is slower than optimized `C` by a factor of two [15]. Recent additions to the compiler infrastructure [16], which provide instruction-level parallelism, will reduce this factor further. As `ADPfusion` is built on top of the `Repa` [17] library for CPU-level parallelism, we can expect improvements in this regard to be available for our dynamic programming algorithms in the near future.

3   Applications

In this section we discuss one elaborate and practically relevant example where a grammar product of two simple grammars yields a complex result grammar. The alignment of two sequences of the same type is typically simplified due to mirrored operations. Recalling the alignment grammar from above, we speak of in/del operations as an insertion in one sequence may just as well be described as a deletion in the other sequence. In addition, it does not matter which sequence is bound to which input tape.

The alignment of a protein sequence to a DNA sequence is, however, more involved. In Fig. 1 we summarize this more elaborate example. The DNA sequence is read in one of three reading frames (RFs), and a deletion or insertion does not yield a "simple" in/del but also a frame shift. This more advanced treatment of DNA characters in triplets is due to the *translation* of DNA into protein in steps of three nucleotides, the "codons" of the genetic code. In Fig. 1 frame shifts (with scoring functions `rf1, rf2`) are allowed only at high cost as they change the transcription of following protein characters completely. Staying within a frame is very cheap, even if this involves the deletion of three characters (`del`).

The protein grammar, on the other hand, has the same simple structure as our previous atomic components of the alignment grammar. Here, we indeed only read a single amino acid, or handle a deletion.

The complexity of the DNA-protein alignment stems from the fact that we need to "align" the different frame shifting possibilities in the DNA input while matching zero to three nucleotides to zero or one amino acid in the protein input. In addition, once a frame shift has occurred all following alignments of three nucleotides against one amino acid are scored in the new reading frame until another frame shift occurs or the alignment is completed.

Our framework simplifies the complexity of designing this algorithm considerably. While the *combined* grammar is highly complex, the individual grammars are rather simple. As already mentioned, the protein "stepping grammar" is one of the simplest possible ones. The DNA grammar is more complex as we need to handle stepping and frame shifts in all three reading frames. But considering that we allow indexed non-terminals and calculations on these indices (modulo 3 in the frame shift case), even the frame shift grammar has only four rules, just twice as much as the simplest stepping grammar.

```
Grammar: DNA
F{i} -> stay <<< F{i}    c c c  Grammar: PRO
F{i} -> rf1  <<< F{i+1} c c     P -> amino <<< P a
F{i} -> rf2  <<< F{i+2} c       P -> del   <<< P
F{i} -> del  <<< F{i}           //                    Product: DnaPro
//                              Grammar: PROdone       DNA >< PRO
Grammar: DNAdone                P -> nil <<< empty     + DNAdone >< PROdone
F{i} -> nil <<< empty           //                     - DNAstand >< PROstand
//                              Grammar: PROstand       //
Grammar: DNAstand               P -> del <<< P
F{i} -> del <<< F{i}            //
//
```

Fig. 1. Atomic grammars for the DNA-protein alignment example. **(I)** Nucleotides are read in triplets (three nucleotides each). The DNA grammar switches between reading frames. DNAdone and DNAstand handle the termining and looping case. **(II)** The PROtein grammar works similarly, but reads only a single amino acid at a time. The expansion of the DNA grammar is more complicated, as the indexed non-terminal symbol F expands to three different non-terminals corresponding to the three possible reading frames. **(III)** The grammar product of DNA and PROtein without the looping case "**stand**" and with the terminating case "**done**". In code, >< represents the direct product ($\otimes$). The resulting 24-production rule grammar is shown in the Supplemental Material together with an extended description.

Together with the grammar an interface (a signature in ADP terms) is generated. This interface simplifies the creation of scoring and backtracking functions. It is here, where the feasibility of certain frame shift operations (via rf1, rf2) is decided. This is advantageous as the grammar describes the full search space while the semantics of DNA-protein alignment are decided solely in the scoring functions.

The resulting 24-production rule grammar is easily calculated in our frame work. We emphasize that it is very easy to extend this grammar to allow for, say, an alignment of two DNA sequences with two protein sequences. This grammar can be *calculated* at basically no additional cost but would pose a daunting task if implemented by hand. An extended description of this grammar, together with a depiction of the 24 production rules can be found in the Supplemental Material[2].

4   Discussion

Summary. We have presented a formal, abstract algebra on linear grammars. This algebra provides operations to create complex, multi-tape grammars from simple, single-tape atomic ones. More informally, we have created a method and implementation to "multiply" dynamic programming algorithms. We also

---

[2] http://www.bioinf.uni-leipzig.de/Software/gramprod/
 hoe-hof-2013-supplement.pdf

provide a compiler framework that makes the grammars readily available for actual deployment with good performance of the resulting code.

The products of linear grammars, despite the simplicity of individual grammars, give rise to many often-used and powerful algorithms where word-like objects are aligned with each other. We have restricted ourselves to a problem from the realm of computational biology, as the alignment of DNA and protein sequences provides a good example of the emerging complexity of algorithmic alignment, especially when the words to be aligned have differing internal structure – in the example case the possibility of a frame shift in the DNA sequence.

Future Work. This work also leads to a number of questions to be answered in the future. We should investigate the actual performance of our automatically generated grammar implementations versus hand-written code, but this is mostly a question delegated to the underlying `ADPfusion` framework. We prefer a separation of concerns: *grammar products* emphasize algebraic operations, the user need not be concerned with low-level implementation details.

We have restricted ourselves to linear grammars, as the next class of formal grammars, context-free grammars, requires us to give a good definition of the direct product on production rules with more than one non-terminal symbols.

The direct product implicitly introduces dependencies that couple the input bands. Consider the product of productions $(X \rightarrow Xa) \otimes (X \rightarrow X)$. There are two possibilities how the right-hand side can be interpreted:

$$\left(\tfrac{X}{X} \right) \rightarrow \left(\tfrac{X}{X} \right)\left(\tfrac{a}{\epsilon} \right) \tag{19}$$

$$\left(\tfrac{X}{X} \right) \rightarrow \left(\tfrac{X}{\epsilon} \right)\left(\tfrac{a}{X} \right) \tag{20}$$

Aligning the two non-terminals to form a new non-terminal as in Equ. (19) is equivalent to a dependence statement. All possible derivations of $\left(\tfrac{X}{X} \right)$ are considered and both tapes are coupled.

The situation is quite different for the production rule given in Equ. (20). Since $\left(\tfrac{X}{\epsilon} \right)$ aligns a substring with the empty string, we basically decouple the two tapes. Furthermore, we formally have constructed a non-terminal $\left(\tfrac{a}{X} \right)$ that "mixes" non-terminals and terminals on different tapes. In Definition 1 we have avoided this complication by restricting ourselves to linear grammars, where constructions akin to Equ. (20) can always be avoided. When attempting to generalize the framework to arbitrary CFGs, however, this is not possible anymore.

Consider, for example the CFG $\mathcal{A} = \{\{S\}, \{x\}, \{S \rightarrow Sx \mid SS\}, S\}$. Even if we give precedence to matching up non-terminals, $\mathcal{A} \otimes \mathcal{A}$ has productions of the form

$$\left(\tfrac{S}{S} \right) \rightarrow \left(\tfrac{SS}{Sx} \right) \simeq \left(\tfrac{S}{S} \right)\left(\tfrac{S}{x} \right) \tag{21}$$

where $\left(\tfrac{S}{x} \right)$ is neither a terminal nor a non-terminal according to Def. 1. One possibility to deal with this issue is to expand the set of non-terminals to

$N = (N_1 \times N_2) \cup (N_1 \times T_2) \cup (T_1 \times N_2)$ and to add productions of the form $\binom{x}{A} \to \binom{x}{\alpha}$ if $x \in T_1$ and $A \to \alpha \in \mathcal{P}_2$ as well as $\binom{A}{x} \to \binom{\alpha}{x}$ if $x \in T_2$ and $A \to \alpha \in \mathcal{P}_1$. The intuition here is that we can have a terminal produced in one factor, while the other factor still presents a non-terminal. Further derivations then can affect only the factor with the non-terminal, while the terminal in the other factor must remain untouched.

A second complication arises e.g. in the following example: $\mathcal{B} = \{\{S\}, \{x\}, \{S \to x \mid SS\}, S\}$. In $\mathcal{B} \times \mathcal{B}$ we now obtain productions of the form $\binom{S}{S} \to \binom{SS}{x}$. A useful resolution in this case is to re-interpret these as

$$\binom{S}{S} \to \binom{SS}{x} \simeq \binom{S}{\epsilon}\binom{S}{x} \mid \binom{S}{x}\binom{S}{\epsilon} \tag{22}$$

i.e., to allow for all alignments of the r.h.s. in the productions. The explicit use of ϵ suggests an alternative extension of terminal and non-terminal symbols sets, respectively. Setting $N = N_1 \times N_2 \cup N_1 \times \{\epsilon\} \cup \{\epsilon\} \times N_2$ and $T = T_1 \times T_2 \cup T_1 \times \{\epsilon\} \cup \{\epsilon\} \times T_2$. In this setting, we would re-interpret the production of Equ. (21) in the following way:

$$\binom{S}{S} \to \binom{SS}{Sx} \simeq \binom{S}{S}\binom{S}{\epsilon}\binom{\epsilon}{x} \mid \binom{S}{S}\binom{\epsilon}{x}\binom{S}{\epsilon} \tag{23}$$

Apart from questions on how to extend algebraic operations on grammars from linear to context-free grammars, we also need to consider scoring algebras for such products. We anticipate that in many cases, a scoring algebra can be expressed as a form of product itself where the two scoring functions (one for each grammar) are themselves combined in some well-defined form. One possibility is the use of a folding operation to combine scores for subsets of the individual dimensions. It then follows that given two algebras $\mathcal{A}_{G_1}$ and $\mathcal{A}_{G_2}$ for grammars G_1 and G_2 we should be able to define an operation $\mathcal{A}_{G_1} \otimes_\tau \mathcal{A}_{G_2}$ which generates appropriate algebras from algebras for atomic grammars. As long as τ has some structure similar to a fold or another operation on subsets of the dimensions (of the grammars) involved, appropriate products can be automatically defined. This becomes especially useful as we want to define ADP-like [18] algebra-products as well, to explore the rich space of combined algebras on grammars constructed from algebraic operations on atomic grammars.

Another avenue of future research is the question of semantic ambiguity of the resulting grammars. Simple products of the same grammar yield ambiguous alignments on sequences of in-dels. This problem is typically dealt with a good grammar design that explicitly allows only one order of successive insertions and deletions on multiple tapes. Automatic dis-ambiguation is probably complicated but would further simplify the creation of complex multi-tape grammars.

Acknowledgements. This work was funded, in part, by the Austrian FWF, project "SFB F43 RNA regulation of the transcriptome". CHzS thanks Jing, Katja, Lydia, and Nancy (and gin, as well as a mad man in a box).

References

1. Lipman, D.J., Altschul, S.F., Kececioglu, J.D.: A tool for multiple sequence alignment. Proc. Natl. Acad. Sci. USA 86(12), 4412–4415 (1989)
2. Giegerich, R., Meyer, C.: Algebraic Dynamic Programming. In: Kirchner, H., Ringeissen, C. (eds.) AMAST 2002. LNCS, vol. 2422, pp. 349–364. Springer, Heidelberg (2002)
3. Giegerich, R., Meyer, C., Steffen, P.: A Discipline of Dynamic Programming over Sequence Data. Science of Computer Programming 51(3), 215–263 (2004)
4. Gotoh, O.: An improved algorithm for matching biological sequences. J. Mol. Biol. 162, 705–708 (1982)
5. Gotoh, O.: Alignment of three biological sequences with an efficient traceback procedure. J. Theor. Biol. 121, 327–337 (1986)
6. Dewey, T.G.: A sequence alignment algorithm with an arbitrary gap penalty function. J. Comp. Biol. 8, 177–190 (2001)
7. Konagurthu, A.S., Whisstock, J., Stuckey, P.J.: Progressive multiple alignment using sequence triplet optimization and three-residue exchange costs. J. Bioinf. and Comp. Biol. 2, 719–745 (2004)
8. Kruspe, M., Stadler, P.F.: Progressive multiple sequence alignments from triplets. BMC Bioinformatics 8, 254 (2007)
9. Steiner, L., Stadler, P.F., Cysouw, M.: A pipeline for computational historical linguistics. Language Dynamics & Change 1, 89–127 (2011)
10. Needleman, S.B., Wunsch, C.D.: A General Method Applicable to the Search for Similarities in the Amino Acid Sequence of Two Proteins. Journal of Molecular Biology 48(3), 443–453 (1970)
11. Sankoff, D.: Simultaneous solution of the RNA folding, alignment and protosequence problems. SIAM Journal on Applied Mathematics, 810–825 (1985)
12. The GHC Team: The Glasgow Haskell Compiler (GHC) (1989–2013), http://www.haskell.org/ghc/
13. Sheard, T., Jones, S.P.: Template Meta-programming for Haskell. In: Proceedings of the 2002 ACM SIGPLAN Workshop on Haskell, pp. 1–16. ACM (2002)
14. Coutts, D., Leshchinskiy, R., Stewart, D.: Stream Fusion: From Lists to Streams to Nothing at All. In: Proceedings of the 12th ACM SIGPLAN International Conference on Functional Programming, ICFP 2007, pp. 315–326. ACM (2007)
15. Höner zu Siederdissen, C.: Sneaking around concatMap: efficient combinators for dynamic programming. In: Proceedings of the 17th ACM SIGPLAN International Conference on Functional Programming, ICFP 2012, pp. 215–226. ACM (2012)
16. Mainland, G., Leshchinskiy, R., Jones, S.P., Marlow, S.: Exploiting vector instructions with generalized stream fusion. In: Proceedings of the 18th ACM SIGPLAN International Conference on Functional Programming (2013)
17. Keller, G., Chakravarty, M.M., Leshchinskiy, R., Peyton Jones, S., Lippmeier, B.: Regular, Shape-polymorphic, Parallel Arrays in Haskell. In: Proceedings of the 15th ACM SIGPLAN International Conference on Functional Programming, ICFP 2010, pp. 261–272. ACM (2010)
18. Steffen, P., Giegerich, R.: Versatile and declarative dynamic programming using pair algebras. BMC Bioinformatics 6(1), 224 (2005)

Influence of Scaffold Stability and Electrostatics on Top7-Based Engineered Helical HIV-1 Epitopes

Isabelle F.T. Viana[1,2,3], Rafael Dhalia[4], Marco A. Krieger[3],
Ernesto T.A. Marques[2,4], and Roberto D. Lins[1]

[1] Department of Fundamental Chemistry, Federal University of Pernambuco,
Recife, PE, 50670-560, Brazil
[2] Center for Vaccine Research and Department of Infectious Diseases and Microbiology,
University of Pittsburgh, PA, 15261, USA
[3] Carlos Chagas Institute, Oswaldo Cruz Foundation, Curitiba, PR, 81350-010, Brazil
[4] Aggeu Magalhães Research Center, Oswaldo Cruz Foundation, Recife, PE, 50670-420, Brazil
roberto.lins@ufpe.br

Abstract. We have recently engineered HIV-1 epitopes into the Top7 protein as scaffold using molecular dynamics simulations. The immunogenicity of the computer-engineered chimeric proteins was verified using human patient sera. The level and quality of the immune response was correlated to the structural stability of the chimeras as determined by molecular dynamics simulations. This work offers support for this correlation by a comparison between the calculated and experimental circular dichroism spectra for a selection of the Top7-HIV-1 chimeric proteins. In addition, analyzes of surface charge distribution suggest that the maintenance of an electrostatic surface potential signature is crucial for the immunogenicity of the *de novo* designed proteins.

1 Introduction

We have previously described the design of 6 chimerical proteins based on the epitope grafting strategy [1]. The amino acid sequences of three putative conformation-specific epitopes from the ectodomain of the HIV-1 gp41 protein were identified (namely R2, R3 and R8) and transplanted into a highly stable scaffold called Top7 (PDB: 1QYS) (Fig 1). Since the three epitopes are know to assume a α-helix conformation on gp41, these sequences replaced one of the helical regions on Top7. The Top7 helices were called helix A (from residue 24 to 41) and helix B (from residue 55 to 72) generating six chimeras: R2HA, R2HB, R3HA, R3HB, R8HA and R8HB (Fig 1).

The designed proteins were tested against HIV-1 positive and negative human serum samples by means of Luminex assays. Only those chimeras able to mimic the conformation pattern found for the same amino acid sequences in the native gp41 protein were able to correctly differentiate between the two sera sample populations (Table 1). The structural properties of these designed proteins were determined by molecular dynamics studies and directly correlated with their immunological reactivity. Based on these findings, we have proposed that the maintenance of the native secondary structure of the grafted epitopes is correlated with their ability of being recognized by their respective antibodies [1].

J.C. Setubal and N.F. Almeida (Eds.): BSB 2013, LNBI 8213, pp. 94–103, 2013.

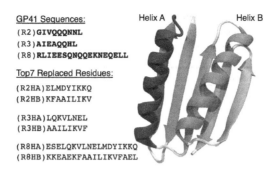

GP41 Sequences:
(R2)**GIVQQQNNL**
(R3)**AIEAQQHL**
(R8)**RLIEESQNQQEKNEQELL**

Top7 Replaced Residues:
(R2HA)ELMDYIKKQ
(R2HB)KFAAILIKV

(R3HA)LQKVLNEL
(R3HB)AAILIKVF

(R8HA)ESELQKVLNELMDYIKKQ
(R8HB)KKEAEKFAAILIKVFAEL

Fig. 1. Cartoon representation of the Top7 scaffold (right) and the lists of three helical GP41 putative epitope sequences (*R2*, *R3* and *R8*) and the replaced residues in Top7 (left). Top7 secondary structure elements are represented as: α-helices: spirals; β-strands: arrows; and loops and unstructured regions: white coils. *Helix A* is shown in red while *helix B* in blue.

The high selectivity and affinity of the antigen-antibody association are consequence of the simultaneous action of several molecular forces exhibiting different magnitudes and distance dependences [2]. These forces are primarily determined by structure complementarity between both molecules and comprise of a range of inter-residue interactions, such as van der Waals, hydrogen bonds, hydrophobic contacts, salt bridges and long-range electrostatic interactions [3].

Molecular modeling, kinetics, and equilibrium measurements have elucidated the relationship between antigen-antibody structural complementarity and the forces that control the association process [2]. Among the involved molecular forces, shape compatibility, hydrophobic interactions and electrostatic complementarity remarkably and directly influence the initial antigen-antibody association and the strength of docked complex [4, 5]. The knowledge that higher electrostatic interactions can make critical contributions to the extent and rate of antibody binding [6] and are correlated with high affinity antibodies has been used to predict putative interaction regions in protein surfaces [7].

Therefore, while secondary structure maintenance must play an important role in the Top7-based HIV-1 antigen-antibody interactions, it is also of interest of this work to evaluate how specific electrostatic distribution can be correlated with antibody recognition by the designed chimerical proteins. Here we utilize a combination of computational and experimental methods to examine the structural and electrostatic properties of these six designed proteins, specifically testing the hypothesis that the maintenance of electrostatic distribution over the epitope surface influences the recognition and interaction with the respective antibodies. The analysis described here not only probe the impact of the structure maintenance on the biological activity of the designed HIV-1 chimeras, but also the impact of the local charge surface pattern on the antibody-antigen recognition.

Table 1. Ability of helix-based chimerical Top7 proteins to discriminate between HIV-1 positive and negative human sera samples[a] and the corresponding molecular structures obtained by molecular dynamics simulations (shown in cartoon model)

Epitope ID	Sensitivity (%)	Specificity (%)	3D Structure from MD[b,c]
R2HA[b]	96	81	
R2HB	57	42	
R3HA[b]	89	67	
R3HB[b]	100	86	
R8HA	53	46	
R8HB	65	53	

[a]Values resulting from the average median fluorescence intensity, as measured by Luminex assays, of each chimera to respond to 26 positive and 21 negative HIV-1 human sera samples; [b]data taken from [1]; [c]helical regions containing the epitope sequence are shown in blue.

2   Methodology

2.1   Computational Methodology

Atomic coordinates for the Top7 protein and chimeras used for the theoretical circular dichroism spectra and electrostatic calculations correspond to the final structures of a 50-ns molecular dynamics simulations previously performed [1]. Starting from the PDB-formatted coordinates, atomic charges and radii have been assigned for all atoms by the PDB2PQR 1.8 Server [8], using the available Amber parameter set. The appropriate protonation state for each tritable residue of every protein was previously determined by the PROPKA 3.0 program [9] and hydrogen atoms added accordingly. Electrostatic potentials were obtained by solving numerically the linear Poisson-Boltzmann equation and applying a finite-difference method [10-12] using the APBS (Adaptive Poisson-Boltzmann Solver) program [13]. A dielectric constant for solvent of 78.54 $C^2/N.m^2$ with solvent radius of 1.4 nm, surface tension of 0.105 N/m, and ionic strength of 0.150 M was used to describe the structures in aqueous solution. The internal dielectric constant of the solute was set to 1 $C^2/N.m^2$ (default value), varying as a function of distance reaching up to 78.4 $C^2/N.m^2$ at the protein solvent-accessible regions. The dielectric coefficient describes the local polarizability. The functional form of this coefficient depends on the molecular shape. A low value (typically between 1 and 20) is usually assigned to the biomolecular core and higher values (ca. 80) are used to represent water at solvent accessible areas. The three-dimensional potentials were obtained using 129 grid points in the x, y and z directions. Protein structures and their corresponding electrostatic potentials were visualized and analyzed with the VMD 1.9 program [14]. The Dichrocalc webserver was used to calculate the theoretical circular dichroism spectra for Top7, R2HA and R3HB proteins from their corresponding atomic coordinates [15]. The averaged variations in the secondary structure content (α-helices and β -sheet) of the chimeras relative to the original Top7 were calculated using the DSSP program [16] over the last 25-ns molecular dynamics simulation window, as recently reported [1].

2.2   Circular Dichroism Spectroscopy

The chimeric proteins were produced in prokaryotic system and purified by affinity chromatography as previously described [1]. The proteins were diluted in 300 mM NaCl, 50 mM Tris-HCl, pH 8.0 buffer (Buffer A) to the concentration of 100 µg/mL and dialyzed in Slide-A-Lyzer Dialysis Cassettes, 10K MWCO (Thermo Scientific). The dialysis procedure was performed during 12 hours at 4 °C against 5 buffers containing decreasing concentrations of NaCl and Tris-HCl as follows: Buffer B – 225 mM NaCl, 50 mM Tris-HCl, pH 8.0; Buffer C – 150 mM NaCl, 40 mM Tris-HCl, pH 8.0; Buffer D – 75 mM NaCl, 30 mM Tris-HCl, pH 8.0; Buffer E – 45 mM NaCl, 25 mM Tris-HCl, pH 8.0; Buffer F – 30 mM NaCl, 20 mM Tris-HCl, pH 8.0. The dialyzed samples were concentrated using Vivaspin 6, 10,000 MWCO (Sartorius Stedim Biotech) and quantified by spectrophotometry. Circular dichroism data were collected on an Olis DSM17 Spectrometer. Far-UV circular dichroism wavelength scans were recorded in a 0.1 cm path length. The proteins[1] were diluted in Buffer F to the following concentrations: Top7_Original – 45.35 μM, Top7_R2HA – 84.12 μM, Top7_R3HB – 93.47 μM. Circular dichroism scans were carried out from 260 to 200 nm with 5 seconds averaging times, 1 nm step size, and 2 nm bandwidth at 25 °C. Spectra were corrected for a buffer blank and baseline molar ellipticity at 260 nm. Scan data were smoothed by the Stavistsky-Golay method [17].

3   Results and Discussion

3.1   Structural Stability

Using the original Top7 scaffold as reference structure, molecular dynamics simulations data (root mean square deviation as a function of simulation time, residue averaged root mean square fluctuations, total number of intra-protein hydrogen bonds, solvent accessible surface area and time evolution of secondary structure motifs) indicated that helical constructs based on R2 and R3 sequences were able to maintain the overall scaffold and secondary structure elements. In contrast, the insertion of R8 sequences in either helix A or B caused heavy loss of α-helical content of the top-7 based chimeras [1].

New trajectory analyses of the relative variation in secondary structure content with respect to the original Top7 is shown in Fig 2. Averages were obtained for the second half of the simulation time (25 to 50 ns period interval) and the differences to the original scaffold calculated. It is worth noting that Top7 has about 30% of its structure in α-helix and that a 10% variation in helical content would represent only about 3 residues in the whole structure. The molecular simulations indicate that grafting of R2 and R3 in Top7 does not alter significantly the helical content of the scaffold. Compared to Top7, R8HA shows a loss of almost 50% of its helical content, *i.e.*, the equivalent of one out of the two helices present in the protein. The loss of helical structure is also pronounced in R8HB, where over 20% of the total helical content is lost. (A depiction of this decline in helical content can be seen by the cartoon representation of final structures from the 50 ns molecular dynamics simulations in Table 1). The absence of significant variations in β-sheet content in all chimeras supports the hypothesis that this motif comprises the core of Top7 structural stability [18, 19]. The larger variation is observed for R3HA, where on average about 8% of β-sheet content is lost.

The partial unfolding observed for R8HA and R8HB chimeras would be an indication of the inability of these proteins to differentiate between HIV-1 positive and negative human sera. In fact, this has been confirmed by Luminex assays [1], also shown here in Table 1 by their calculated average sensibility and specificity to the human sera samples used. On the other hand, the R2HB chimera presents a poor immunological performance (Table 1), similar to the partially unfolded R8HA and R8HB chimeras. We have previously attributed it to a potentially intrinsic flexibility of the grafted epitope in the scaffold. However, the data presented on Fig. 2 shows a rather negligible variation in secondary structure content for this chimera.

---

[1] Protein selection was based on sample availability.

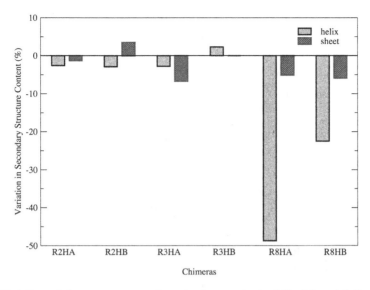

Fig. 2. Variation of the average secondary structure content of Top7-based helical HIV-1 epitope chimeras over a 25 to 50 ns period from previously performed molecular dynamics simulations [1]. The Top7 protein is taken as reference for secondary structure content.

Aiming to validate the theoretical predictions of secondary structure content (at least partially), we have acquired experimental circular dichroism spectra for two helical chimeras (R2HA and R3HB) and the original Top7 scaffold and compared to the corresponding theoretically calculated spectra from the atomic coordinates of each protein (Fig. 3). Circular dichroism spectroscopy measures differences in the absorption of polarized light (left-handed versus right-handed) that arise from structural asymmetry. Therefore, the secondary structure of proteins can be probed due to the asymmetry of the peptide bonds. The amide chromophore largely dominates the spectra of proteins at far-UV. Amides have two well-characterized electronic transitions of low energy, n$\rightarrow$π* and π$_0$$\rightarrow$π* responsible to produce signals at the 215 to 230 nm and 185 to 200 nm intervals, respectively [20, 21]. Our measurements were carried out from 200 nm to 260 nm, therefore special attention is given to agreement between theoretical and calculated spectra at the 208 nm (π$_0$$\rightarrow$π* transition) and 222 nm (n$\rightarrow$π* transition) corresponding to α-helix signals, and bands in the region between 216 and 218 nm (n$\rightarrow$π* transition) that are typical of β-sheet content.

Fig. 3 shows a remarkable agreement between theoretical and experimental spectra for all three proteins from 205 to 260 nm interval. This result indicates that the structures obtained by molecular dynamics simulations for Top7, R2HA and R3HB were able to appropriately describe the secondary structure content and consequently representative conformations of these proteins in solution. Furthermore, it validates the assessment of secondary structure content for the remaining systems and ensures the correlation between protein scaffold maintenance and appropriate immunological activity. On the other hand, such validation leaves the poor immunoreactivity performance of the stable R2HB as an open question.

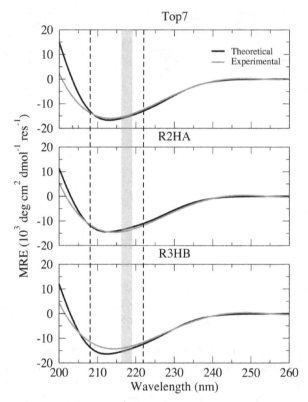

Fig. 3. Comparison of theoretical and experimental circular dichroism spectra of original Top7, R2HA and R3HB proteins. Dashed lines mark 208 nm and 222 nm, points in the spectra corresponding to helical content, while the region from 216 to 218 nm corresponding to β-sheet content is highlighted by a transparent band. (MRE stands for mean residue ellipicity).

3.2   Epitope Electrostatic Surface Potential

Theoretical and experimental data indicates that maintenance of the epitope native conformation when grafted onto a Top7 scaffold is directly correlated to the ability to be recognized by an antibody and therefore likely to elicit an immunological response. However, the R2HB chimera seems to be an exception to the rule. While the stability of the secondary structure content of the protein is comparable to other immunoreactive chimeras, its performance to differentiate between HIV-1 positive and negative human sera samples renders this protein unacceptable to be considered even as a lead for diagnostics or vaccines (see Table 1).

Protein-protein interactions require shape (structural motif) and electrostatic complementarity [5, 22-24]. This general principle is also valid for antigen-antibody interactions, as largely showcased in the literature. The electrostatic patterns exhibited by epitopes has been considered crucial for recognition [25] and sometimes the major contributor to binding energy [26]. Despite of its importance, the electrostatic pattern of a protein motif can be dramatically affected by its neighboring residues. Changes in the electrostatic properties responsible to inactivate an antigen can be caused by point mutations of residues being spatially as far as 9 Å [27]. Therefore, the surface potential of an epitope grafted onto a scaffold warrants consideration.

In light of this information, we ponder whether residues in the Top7 scaffold may affect the charge distribution of grafted epitopes. This question has been addressed by examining the electrostatic potential on the molecular surface of the stable chimeric antigens R2HA, R2HB, R3HA and R3HB. Fig 4 shows a common electrostatic signature for the R3 sequence whether it replaces residues in helix A or helix B in Top7. However, a quite different electrostatic pattern is shown for the R2 sequence depending on the helix it has been exhibited.

Comparison of the electrostatic potential of these regions (R2 and R3 sequences) within the gp41 protein is not possible since a complete packed structure is not currently available. Nevertheless, it is reasonable to assume that a common surface charge distribution for both R3-based epitopes is associated to the fact that both proteins displayed high levels of immunoreactivity against HIV-1 human positive sera samples. In other words, similarity of the electrostatic potential would translate into a conserved of three-dimensional binding pattern. Such support comes from a visual inspection of the surface charge distribution of R2HA. If rotated by 180°, a similar pattern to the ones displayed by R3HA and R3HB can be seen (Fig 4). Despite the structural stability of R2HA, this epitope when featured on helix A of Top7 does not show a common electrostatic potential on its surface. This result indicates that residues in the Top7 scaffold might affect the electrostatic properties of grafted sequences and therefore such analysis should be taken into account when designing biologically relevant Top7-based heterologous sequences.

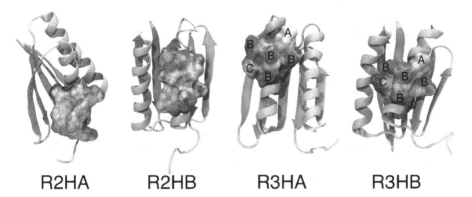

R2HA   R2HB   R3HA   R3HB

Fig. 4. Surface charge of the helical HIV-1 epitopes grafted onto the structurally stable Top7-based scaffolds. Proteins are shown in cartoon model. Surface charge is represented by the corresponding electrostatic potential mapped onto the molecular surface of the epitope region. Positively charged potential is represented in blue, negatively charged potential in red and neutral regions in white. Potential interval is from -5 to +5 $kJ.mol^{-1}.e^{-1}$. The high similarity of the electrostatic potentials for the R3 sequence in helices A and B is highlighted by assigning labels over the displayed potential, where *A* is placed over an apolar region, *B* over a negatively charged region and *C* over a positively charged area of the proteins.

4   Conclusion

Comparison of theoretical and experimental circular dichroism spectra for selected systems was used to validate the structural data for the chimeras obtained from our previous molecular dynamics simulations. Relative variation in secondary structure content and electrostatic calculations revealed that recognition of the Top7-based HIV-1 chimeras by the antibody depends on both: i) the maintenance of the structural stability of the scaffold and epitope as close as to its native conformation, and ii) the charge distribution pattern on the epitope surface to ensure shape and electrostatic complementarity. Furthermore, these results contribute to better understand the forces that control the antigen-antibody interaction and guide the rational design of high affinity proteins for a variety of engineered functions.

Acknowledgments. This work was supported by FACEPE, NanoBiotec-BR/CAPES, CNPq, INCT-INAMI, nBioNet and the Oswaldo Cruz Foundation.

References

1. Viana, I.F.T., Soares, T.A., Lima, L.F.O., Marques, E.T.A., Krieger, M., Dhalia, R., Lins, R.D.: De Novo Design of Immunoreactive Conformation-specific HIV-1 Epitopes based on Top7 Scaffold. RSC Advances 3, 11790–11800 (2013)
2. Leckband, D.E., Kuhl, T.L., Wang, H.K., Muller, W., Herron, J., Ringsdorf, H.: Force probe measurements of antibody-antigen interactions. Methods 20, 329–340 (2000)
3. Janeway Jr., C.A., Travers, P., Walport, M., et al.: Immunobiology: The Immune System in Health and Disease. Garland Science, New York (2001)
4. Tsai, C.J., Lin, S.L., Wolfson, H.J., Nussinov, R.: Studies of protein-protein interfaces: a statistical analysis of the hydrophobic effect. Protein Sci. 6, 53–64 (1997)
5. McCoy, A.J., Chandana Epa, V., Colman, P.M.: Electrostatic complementarity at protein/protein interfaces. Journal of Molecular Biology 268, 570–584 (1997)
6. Sinha, N., Mohan, S., Lipschultz, C.A., Smith-Gill, S.J.: Differences in electrostatic properties at antibody-antigen binding sites: implications for specificity and cross-reactivity. Biophys. J. 83, 2946–2968 (2002)
7. Fiorucci, S., Zacharias, M.: Prediction of protein-protein interaction sites using electrostatic desolvation profiles. Biophys. J. 98, 1921–1930 (2010)
8. Dolinsky, T.J., Czodrowski, P., Li, H., Nielsen, J.E., Jensen, J.H., Klebe, G., Baker, N.A.: PDB2PQR: expanding and upgrading automated preparation of biomolecular structures for molecular simulations. Nucleic Acids Research 35, W522–W525 (2007)
9. Rostkowski, M., Olsson, M.H.M., Sondergaard, C.R., Jensen, J.H.: Graphical analysis of pH-dependent properties of proteins predicted using PROPKA. BMC Structural Biology 11, 6 (2011)
10. Davis, M.E., McCammon, J.A.: Calculating Electrostatic Forces from Grid-Calculated Potentials. Journal of Computational Chemistry 11, 401–409 (1990)
11. Nicholls, A., Honig, B.: A Rapid Finite-Difference Algorithm, Utilizing Successive over-Relaxation to Solve the Poisson-Boltzmann Equation. Journal of Computational Chemistry 12, 435–445 (1991)
12. Antosiewicz, J., Gilson, M.K., McCammon, J.A.: Acetylcholinesterase - Effects of Ionic-Strength and Dimerization on the Rate Constants. Israel Journal of Chemistry 34, 151–158 (1994)

13. Baker, N.A., Sept, D., Joseph, S., Holst, M.J., McCammon, J.A.: Electrostatics of nanosystems: Application to microtubules and the ribosome. Proceedings of the National Academy of Sciences of the United States of America 98, 10037–10041 (2001)
14. Humphrey, W., Dalke, A., Schulten, K.: VMD: visual molecular dynamics. J. Mol. Graph. 14, 33–38 (1996)
15. Bulheller, B.M., Hirst, J.D.: DichroCalc-circular and linear dichroism online. Bioinformatics 25, 539–540 (2009)
16. Kabsch, W., Sander, C.: Dictionary of protein secondary structure: pattern recognition of hydrogen-bonded and geometrical features. Biopolymers 22, 2577–2637 (1983)
17. Savitzky, A., Golay, M.J.E.: Smoothing and Differentiation of Data by Simplified Least Squares Procedures. Anal. Chem. 36, 1627–1639 (1964)
18. Boschek, C.B., Apiyo, D.O., Soares, T.A., Engelmann, H.E., Pefaur, N.B., Straatsma, T.P., Baird, C.L.: Engineering an ultra-stable affinity reagent based on Top7. Protein Engineering, Design & Selection 22, 325–332 (2009)
19. Soares, T.A., Boschek, C.B., Apiyo, D., Baird, C., Straatsma, T.P.: Molecular basis of the structural stability of a Top7-based scaffold at extreme pH and temperature conditions. J. Mol. Graph. Model. 28, 755–765 (2010)
20. Woody, R.W., Koslowski, A.: Recent developments in the electronic spectroscopy of amides and alpha-helical polypeptides. Biophys. Chem. 101-102, 535–551 (2002)
21. Correa, D.H., Ramos, C.H.I.: The use of circular dichroism spectroscopy to study protein folding, form and function. African J. Biochem. Res. 3, 164–173 (2009)
22. Sinha, N., Smith-Gill, S.J.: Electrostatics in protein binding and function. Curr. Protein Pept. Sci. 3, 601–614 (2002)
23. Jones, S., Thornton, J.M.: Principles of protein-protein interactions. Proc. Natl. Acad. Sci. U. S. A. 93, 13–20 (1996)
24. Janin, J.: Elusive affinities. Proteins 21, 30–39 (1995)
25. Gabb, H.A., Jackson, R.M., Sternberg, M.J.E.: Modelling protein docking using shape complementarity, electrostatics and biochemical information. Journal of Molecular Biology 272, 106–120 (1997)
26. Sinha, N., Mohan, S., Lipschultz, C.A., Smith-Gill, S.J.: Differences in Electrostatic Properties at Antibody-Antigen Binding Sites: Implications for Specificity and Cross-Reactivity. Biophysical Journal 83, 2946–2968 (2002)
27. Chien, N.C., Roberts, V.A., Giusti, A.M., Scharff, M.D., Getzoff, E.D.: Significant structural and functional change of an antigen-binding site by a distant amino acid substitution: proposal of a structural mechanism. Proc. Natl. Acad. Sci. U. S. A. 86, 5532–5536 (1989)

Random Forest and Gene Networks
for Association of SNPs to Alzheimer's Disease

Gilderlanio S. Araújo[1], Manuela R.B. Souza[3,4], João Ricardo M. Oliveira[3,4], Ivan G. Costa[1,2], and for the Alzheimer's Disease Neuroimaging Initiative*

[1] Center of Informatics - Federal University of Pernambuco, Recife-PE, Brazil
[2] Interdiciplinary Center for Clinical Research (IZKF) and Institute for Biomedical Engineering, RWTH University Hospital, Aachen, Germany
[3] Keizo Asami Laboratory (LIKA) Federal University of Pernambuco, Recife-PE
[4] Neuropsychiatry Department - Federal University of Pernambuco, Recife-PE, Brazil
{gsa3,igcf}@cin.ufpe.br, joao.ricardo@ufpe.br, manu.brsouza@gmail.com
www.cin.ufpe.br

Abstract. Machine learning methods, such as Random Forest (RF), have been used to predict disease risk and select a set of single nucleotide polymorphisms (SNPs) associated to the disease on Genome-Wide Association Studies (GWAS). In this study, we extracted information from biological networks for selecting candidate SNPs to be used by RF, for predicting and ranking SNPs by importance measures. From an initial set of genes already related to a disease, we used the tool GeneMANIA for constructing gene interaction networks to find novel genes that might be associated with Alzheimer's Disease (AD). Therefore, it is possible to extract a small number of SNPs making the application of RF feasible. The experiments conducted in this study focus on investigating which SNPs may influence the susceptibility to AD.

Keywords: Random Forest, SNP, Alzheimer's Disease, Genome-wide Association Study.

1 Introduction

Genome-Wide Association Studies are becoming widespread given the reducing costs of large scale genotyping techniques, such as SNP arrays [6]. A major task for GWAS is to establish and discover new loci and biomarkers for understanding the progress of neurodegeneration caused by AD. This trait is the most common progressive neurodegenerative disease and represents a major cause of loss of neuronal functions followed by cognitive impairments and memory deficiencies of

---

* Data used in preparation of this article were obtained from the Alzheimers Disease Neuroimaging Initiative (ADNI) database (adni.loni.ucla.edu). As such, the investigators within the ADNI contributed to the design and implementation of ADNI and/or provided data Rev March 26, 2012 but did not participate in analysis or writing of this report. A complete listing of ADNI investigators can be found at: http://adni.loni.ucla.edu/wpcontent/uploads/how_to_apply/ADNI_Acknowledgement_List.pdf

J.C. Setubal and N.F. Almeida (Eds.): BSB 2013, LNBI 8213, pp. 104–115, 2013.
© Springer International Publishing Switzerland 2013

elderly patients. AD can be divided into three phases, an initial pre-symptomatic phase, a prodromal stage known as Mild Cognitive Impairment (MCI) and a third phase when patients show dementia with impairments in multiple domains and loss of function in daily activities [1].

Several Alzheimer's GWAS have been performed and associated SNPs have been catalogued on the AlzGene database [3]. However, the tasks of association of variants to phenotypes or the determination of disease from variants still present computational and statistical challenges given the dimensionally of the problem. Usual GWAS data sets have hundreds of thousands of SNPs over hundreds of patients. Traditional univariate statistical methods cannot detect epistatic effects and suffer from multiple testing problems [16]. Therefore, they miss polygenic interactions common in complex diseases and small marginal effects [1]. Additionally, another common property of genetic variants is the fact that some alleles occur together more frequently than expected by chance. This non-random paired combination between SNPs is referred to as linkage disequilibrium (LD) and makes SNP data to be correlated.

Recent works propose the application of random forests (RF) to select SNPs in genome-wide data sets to identify potential risk variants for complex diseases [16, 10–12, 17, 19]. RF is suited to large dimensional data sets, can capture high order and non-linear SNP effects and presents a simple interpretation [11]. However, RF performance and generalization power is restricted in data sets with extremely high dimension such as GWAS data. The common approach to solve this is the use of standard association analysis filters to greatly reduce the number of candidate SNPs previous to RF application [17]. Our attempt to replicate such strategies in data from the Alzheimers Disease Neuroimaging Initiative (ADNI) failed. Random forest showed random predictive errors, possibly because of the small sample size of the data set and lack of biological relevance of the SNPs selected by the used filters[20]. Moreover, the presence of covariance between SNP, as an effect of the linkage disequilibrium, further deteriorates RF performance [11, 12].

As a solution, we propose the use of biological driven selection of SNPs as input to RF to identify potential causal variants for complex diseases [8]. For this, we use genes and SNPs previously reported in meta-analysis provided by AlzGene database to grow our set of candidate SNPs with evidence from gene networks [3]. The network analysis is based on the method GeneMANIA, which uses biological networks from data sources such as: gene expression, protein-protein interaction, physical interaction and shared protein domains [7]. We are particularly interested in genetic risk factors that comprise the architectural base of MCI and AD for two experimental case-control studies provided by the ADNI. As far as we are concern, this is the first work exploring the combination of RF and gene interaction networks to analyse the data from ADNI.

This work is organized as follow. First, we will give a brief explanation of the data and methodology in Section 2. Next, we evaluate the performance of the gene network based selection of SNPs and Random Forest in predicting the patient risk (Section 3). In Section 4, we will investigate the top ranked

SNPs as indicated by Random Forest, evaluate effects of LD in the SNP ranking and compare the results with the classical univariate method Fisher's Exact test. Lastly, we discuss two interesting biological pathways related to Alzheimer: lipids and endocytosis; and analyse the ability of Random Forest or the Fisher's Exact test in recovering genes from such pathways (Section 5).

2   Material and Methods

2.1   SNP Genotyping and Cohort Information

The ADNI efforts are concentrated on discovery of relevant genetic markers for AD and MCI [1]. The individuals were genotyped using the Human 610-Quad BeadChip manufactured by Illumina Inc. More information about the protocol can be found in [4] and clinical categorization in [5]. All genomic coordinates were based on the Genome build 36.2. The ADNI genotype subset used in this paper is divided as follows: 205 humans controls (HC), these individuals are not affected by cognition or neurology disorders; 330 individuals present MCI and 169 individuals developed AD. So, among all SNPs, we restricted for those labeled with 'rs' identifiers, cataloged on dbSNP.

2.2   Random Forest

The RF model is a collection of CART trees for regression or classification problems [15]. RF makes use of bootstrap samples from original data sets for generating a number of trees. Samples not included in the training data, called 'out-of-bag'

---

[1] Data used in the preparation of this article were obtained from the Alzheimers Disease Neuroimaging Initiative (ADNI) database (adni.loni.ucla.edu). The ADNI was launched in 2003 by the National Institute on Aging (NIA), the National Institute of Biomedical Imaging and Bioengineering (NIBIB), the Food and Drug Administration (FDA), private pharmaceutical companies and non-profit organizations, as a $60 million, 5-year public-private partnership. The primary goal of ADNI has been to test whether serial magnetic resonance imaging (MRI), positron emission tomography (PET), other biological markers, and clinical and neuropsychological assessment can be combined to measure the progression of mild cognitive impairment (MCI) and early Alzheimers disease (AD). Determination of sensitive and specific markers of very early AD progression is intended to aid researchers and clinicians to develop new treatments and monitor their effectiveness, as well as lessen the time and cost of clinical trials. The Principal Investigator of this initiative is Michael W. Weiner, MD, VA Medical Center and University of California San Francisco. ADNI is the result of efforts of many co-investigators from a broad range of academic institutions and private corporations, and subjects have been recruited from over 50 sites across the U.S. and Canada. The initial goal of ADNI was to recruit 800 adults, ages 55 to 90, to participate in the research, approximately 200 cognitively normal older individuals to be followed for 3 years, 400 people with MCI to be followed for 3 years and 200 people with early AD to be followed for 2 years. For up-to-date information, see www.adni-info.org.

(OOB), are used to evaluate the prediction performance by the 'out-of-bag error' (EOOB). After building all classification trees, a majority vote method is used to account for the number of votes and the decision is that with most votes. Only a subset of variables is used to grow CART trees, so RF is based on weak learners. The number of trees ($ntree$) and number of variables ($mtry$) parameters has great importance, since increasing the amount and size of trees improves classification performance at the expense of the computational cost. The definition of parameter values is crucial for generalization of the RF method. Lastly, RF allow ranking of variables by the use of the variable importance measures. Here, we use the total decrease in node impurities from splitting on the variable averaged over all trees, where the node impurity is measured by the Gini index [15].

2.3   GeneMANIA: Construction of Heterogeneous Gene Networks

GeneMANIA [7] is a service for integrating gene networks constructed from heterogeneous data and for predicting gene function in real time. GeneMANIA has two components: a) a heuristic algorithm derived from a ridge regression to integrate multiple networks and b) a functional association algorithm to predict function of gene from a network through a process of label propagation. GeneMANIA bases its prediction on data from a wide range of data sources, such as BIOGRID, Pathway Commons and public highthrouput data [29, 26] for constructing six categories of knowledge based networks: co-expression, co-localization, genetic interaction, physical interaction, predicted and shared protein domain interaction networks. The method takes a list of gene labels to identify the most related genes using a guilt-by-association approach. GeneMANIA assigns adaptive weights to the different connections to assess how well genes are connected.

2.4   Methodology Overview

First, we filter SNPs which has more than 10% of missing genotypes. Next, all SNPs were tested by Hardy-Weinberg Equilibrium [30]. All SNPs were kept as no SNPs failed the test (p-$value$ < 0.01). Additionally, we filter SNPs in LD by r^2 metric [30] and we remove those with linkage disequilibrium ($r^2 > 0.7$) with SNPs in the AlzGene database. The idea is to exclude SNPs that are in LD with those already catalogue in the Alzgene database. All these procedures were implemented on PLINK [13].

   Next, we use GeneMANIA as a source of biological knowledge for prioritising SNPs. We extract a set of known genes related to AD based on meta-analyses of GWAS catalogued on AlzGene database [3]. The query genes are the MS4A4E (membrane-spanning 4-domains, subfamily A, member 4E), CR1 (complement component (3b/4b) receptor 1), CLU (clusterin), ABCA7 (ATP-binding cassette, sub-family A (ABC1), member 7), BIN1 (bridging integrator 1), PICALM (phosphatidylinositol binding clathrin assembly protein), APOE (apolipoprotein E), CD33 (CD33 molecule), CD2AP (CD2-associated protein) and MS4A6A (membrane-spanning 4-domains, subfamily A, member 6A). We used the label

ENSG00000214787, as an alternative to the gene MS4A6A, as GeneMANIA did not recognize the primary nomenclature. These genes contain the most significant SNPs associated to the disease so far.

The use of biological knowledge to prioritize variants can build analytical models with more biological sense [8]. So, we constructed three gene networks using the AlzGene genes as a seed list by selecting networks with 50, 100 and 200 neighbouring genes. SNPs were assigned to genes if they were located inside the gene's coding region. After construction of gene networks, we built three gene sets based on information reported by AlzGene. The 'related' set have genes/SNPs present in the AlzGene database, 'not-related' set have genes/SNPs not present in the AlzGene database. We also construct a 'combined' gene set containing SNPs from the previous two sets. This allows us to compare the contribution of genes related and non-related with AD on prediction of an individual's phenotype. A simple summary is shown in Table 1. Note that genes have usually many SNPs associated and that genes from the neighbouring network already have been catalogues in Alzgene and are therefore considered to belong to the related set.

We use the RF implementation provided in R [15, 14]. For RF prediction experiments, we assigned 50, 100, 250, 500 and 1000 to the number of trees parameter and we keep the default number of variables to split per node, so $\sqrt{p}$ where p is the total number of variables. These parameters choices were based on indications from previous work [15]. Thus, we desire to find the lowest values for both parameters that lead to the lowest classification error, for data sets defined in Table 1.

Table 1. Summary of datasets used for prediction and feature selection tasks. This table also presents network data and the number of genes and SNPs as the final variants for analysis.

Gene Networks			Genes : SNPs (p)		
Experiment	Design	Network Size	Related	Not related	Combined
AD-50	HC vs. AD	50	14 : 241	37 : 525	51 : 766
AD-100	HC vs. AD	100	18 : 428	70 : 1.219	98 : 1.647
AD-200	HC vs. AD	200	29 : 596	164 : 3.792	193 : 4.387
MCI-50	HC vs. MCI	50	14 : 241	37 : 525	51 : 766
MCI-100	HC vs. MCI	100	18 : 428	70 : 1.219	98 : 1.647
MCI-200	HC vs. MCI	200	29 : 596	164 : 3.792	193 : 4.387

3   Prediction of Alzheimer Risk by Random Forest

We explored the RF prediction for datasets presented on Table 1. Thus, the performance of the prediction method RF can be compared from different perspectives. The EOOB predictions were summarized in Figure 1. In both control-case studies, there were a slight improvement in EOOB regarding the baseline values of error (0.36 for HC-MCI and 0.45 for HC-AD). For larger networks (100 and

200) we observe a decrease in EOOB for higher number of trees. Such results are better than the ones reported with the use of classical association filters [20]. Overall, non-related and combined sets have similar or better performances than the disease related SNPs. This fact indicates that these non-related SNPs can discriminate MCI and AD as well as known SNPs. We reinforce that the main task in GWAS is indeed the selection of novel relevant markers. These results indicates that SNPs from the non-related/combined sets are potential novel candidates for Alzheimer.

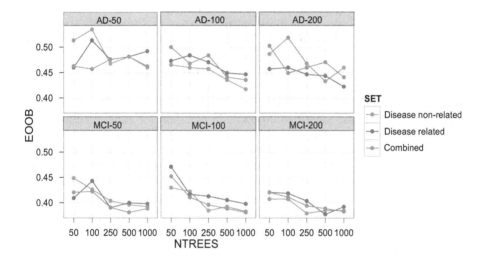

Fig. 1. Estimation of EOOB for subsets presented on Table 1. We consider only a default value $\sqrt{p}$ for *mtry* parameter. In the header of each graph presents the experimental design and gene network to filter SNPs.

4   Ranking of SNPs

4.1   LD and SNP Selection

As case studies for the SNP selection, we use the experiments AD-100 and MCI-100 (see Table 1). Previous work has reported that SNPs in LD might negatively influence the selection of SNPs by RF [11, 12, 20]. Therefore, we investigated the impact of LD on the RF ranking based on the Gini impurity index. For the original data set, we filtered all SNPs with high LD ($r^2 > 0.7$). The two data sets had initially 1647 SNPs and after filtering 1081 (HC-AD) and 1089 (HC-MCI) SNPs.

To analyse impact effects of LD in GI index, we extracted the importance values of 50 runs of the RF method with parameters *ntree* to 1000 tress and *mtry* assigned to $\sqrt{p}$, where p is the number of SNPs. As seen if Figure 2 (A), there is an increased value of average importance at the two experimental designs after filtering of SNPs in LD. Such differences also impact importance *ranking* of SNPs, as observed in [11, 12]. We also investigate the impact of LD filtering on RF prediction (Figure 2B). There is a slight decrease in EOOB for the HC-AD and HC-MCI data sets.

4.2   Comparison with Univariate Methods

The final set of candidate SNPs from Random Forests was the top 50 SNPs regarding the Gini index. Moreover, we apply Fisher's Exact (FE) to these 50 SNPs and accept all SNPs with a p-value of 0.05. Note that the use of only 50 SNPs, which greatly reduces the effects of multiple testing and makes the Fisher's Exact test results to be more optimistic than in a real application. The rankings for Random Forest and p-values for Fischer Exact Test are presented in Table 2. We also report in the column 'Annotation', whenever the SNP was is the AlzGene data base (A) or is related to either the two biological pathways of interests: lipids (L) or endocytosis (E). See the next section for the biological discussion of relevant lipids and endocytosis genes.

RF and Fisher's test identified 40 SNPs in common for experimental design AD-100, while for the MCI-100 30 SNPs were in common. Out of the 50 SNPs from AD-100, 16 were already catalogued in AlzGene of which 12 passed the Fisher's Exact test. For MCI-100, 14 SNPs were catalogued in AlzGene of which only 4 passed the Fisher's test. This indicates that the RF ranking has novel potential SNPs not captured by the FE test. Note that SNPs with low FE p-values (RIMBP2, LR2RA,SNAP) are also top ranked by RF. These indicate that both methods are capturing SNPs with main effects, while RF also captures secondary effects.

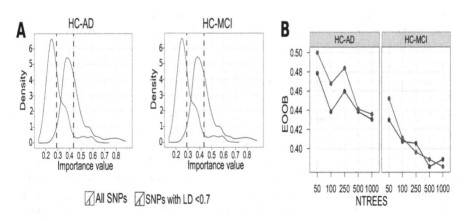

Fig. 2. **(A)** Density of importance values for SNPs for the final experiments HC-100 and MCI-100. The density is plotted considering the combined SNP set before and after LD filtering. **(B)** EOOB estimation after LD pruning and prioritizing SNPs by gene.

5   Gene Analysis

After identification of SNPs and consequently correspondent genes for all SNPs, we identify genes belonging to two biological pathways relaved to Alzheimer.

5.1   Lipids and Alzheimer's Disease

The brain is rich in cholesterol. This constitutes myelin and glial on neuronal membranes. There is abundant evidence that abnormalities in cholesterol metabolism is a risk factor in the pathogenesis of AD. Myelin is a component rich in cholesterol, which accelerates the connections between neurons. There is no degradation of cholesterol in the brain, so excess is converted to 24S-hydroxycholesterol which can be easily transported by lipoproteins that circulate

Table 2. Ranking of SNPs for the Case-Control Design HC-AD (left) and MCI-AD (right). Abbreviations: F, Identified by Fisher's Exact test: L, lipids; E, endocytosis; A, Deposited on AlzGene; - no association.

Rank	Gene	SNP	Annotation	FE p-$value$	Rank	Gene	SNP	Annotation	FE p-$value$
1	SYNJ2	rs10806791	E	0.001	1	IL2RA	rs3134883	-	0.0009
2	RIMBP2	rs7315262	-	0.0008	2	SYNJ1	rs1783099	E	0.03
3	IL2RA	rs7072398	-	0.001	3	KIRREL	rs2777819	-	0.01
4	DNMBP	rs10883428	A	0.001	4	LDLRAP1	rs6687605	L	0.05
5	IL2RA	rs706779	-	0.005	5	SYNJ2	rs10806791	E	0.01
6	DNMBP	rs7089178	A	0.01	6	NCF4	rs909484	-	0.06
7	IL2RA	rs7917726	E	0.005	7	PSTPIP1	rs4078354	-	0.04
8	SNAP91	rs217323	E	0.007	8	LASP1	rs226229	-	0.01
9	DNMBP	rs10883422	A	0.01	9	SH3GL2	rs3824370	E/L	0.04
10	RIMBP2	rs10848114	-	0.009	10	AMPH	rs6962805	E	0.01
11	C7	rs2455314	A	0.008	11	LDLR	rs5930	A/L	0.01
12	HIP1	rs2240133	A/L	0.008	12	CD33	rs1354106	A	0.03
13	RIMBP2	rs756186	-	0.02	13	NEBL	rs12777530	-	0.03
14	AMPH	rs17500182	E	0.002	14	DNMBP	rs7089178	A	0.02
15	CD2AP	rs1385741	A	0.01	15	CD2AP	rs9296562	A	0.03
16	CD2AP	rs9296562	A	0.01	16	HIP1	rs1167797	E/L	0.05
17	CBL	rs7946919	E	0.01	17	BIN1	rs934826	A	0.07
18	DNMBP	rs11190305	A	0.01	18	SH3RF2	rs6869382	-	0.04
19	RIMBP2	rs11060872	-	0.04	19	DNMBP	rs10883422	A	0.2
20	ABCA7	rs3752237	A	0.03	20	NPHS2	rs3765548	-	0.005
21	SH3GL2	rs2208494	E	0.02	21	VEGFA	rs3025010	-	0.04
22	IL2RA	rs3134883	-	0.01	22	NEBL	rs3858202	-	0.04
23	PICALM	rs7938033	A/E	0.2	23	IL2RA	rs791589	-	0.02
24	AMPH	rs6962805	E	0.1	24	FYN	rs1465061	A	0.03
25	MMP25	rs10431961	-	0.02	25	IL2RA	rs791587	-	0.05
26	SYNJ2	rs9459056	E	0.04	26	BIN1	rs10194375	A	0.06
27	IL2RA	rs791587	-	0.03	27	CD2AP	rs3818866	A	0.09
28	NEBL	rs7074881	-	0.02	28	CR1	rs12034383	A	0.4
29	HIP1	rs4588797	E/L	0.03	29	NEBL	rs661924	-	0.07
30	LDLRAP1	rs6687605	E/L	0.9	30	SH3GL2	rs3808750	E/L	0.04
31	ABCA1	rs4149279	A/L	0.01	31	KIRREL	rs3856266	-	0.09
32	LDLR	rs5930	A/L	0.1	32	SH3RF2	rs6898375	-	0.07
33	SH3GL2	rs3824370	E/L	0.1	33	CD33	rs33978622	A	0.04
34	DNMBP	rs10883432	A	0.01	34	HIP1	rs1167830	E/L	0.2
35	RIMBP2	rs10773780	-	0.03	35	BIN1	rs749008	A	0.04
36	KIRREL	rs12049103	-	0.02	36	NPHS2	rs745317	-	0.02
37	SH3PXD2B	rs2569218	E	0.03	37	IL2RA	rs11256497	-	0.04
38	AMPH	rs4720279	E	0.1	38	CBL	rs11217191	E	0.1
39	ENTHD1	rs6001678	-	0.05	39	CBL	rs7946919	E	0.05
40	AMPH	rs2299945	E	0.03	40	PICALM	rs664629	A/E	0.4
41	DNMBP	rs7078153	A	0.01	41	LIPC	rs17190650	A/L	0.1
42	CBL	rs11217191	E	0.07	42	NPHS2	rs2274623	-	0.1
43	SYNJ2	rs7768038	E	0.03	43	CR1	rs10779339	A	0.1
44	MS4A6A	rs17602572	A	0.05	44	SH3GL2	rs3780247	E/L	0.03
45	DNMBP	rs12415442	A	0.03	45	SH3GL3	rs1896799	E	0.07
46	SH3GL2	rs3808750	E	0.04	46	NPHS2	rs2274625	-	0.06
47	MS4A6A	rs662196	A	0.03	47	DNMBP	rs11190305	A	0.07
48	CD2AP	rs3818866	A	0.2	48	NEBL	rs638929	-	0.04
49	DNMBP	rs4919402	A	0.08	49	MS4A4A	rs4939331	A	0.04
50	LASP1	rs226229	-	0.2	50	CD2AP	rs1385741	A	0.06

in the cerebrospinal fluid. The *ABCA1* gene product regulates the homeostasis of lipids and lipoproteins in the central nervous system. The *LDLR* previously associated with AD, suggests functional interactions in conjunction with *APOE*. Genes family members of *LDLR* are expressed in neuronal cells and mediate endocytosis processes in interaction with *APOE* and *APP* genes. In SNP rankings we identified for AD genes as *LDLRAP1, PICALM, HIP1* and *SH3PXD2B*. These performs functions including phospholipid binding and phosphatidylinositol binding, besides the transport of cholesterol to be played by genes *APOE, CLU* and *LDLRAP1*. The *LDLRAP1* is the *LDLR* gene family and is related to receptor binding particles of low density lipoproteins and regulation of cholesterol metabolic process. Based on the evidence gathered by GeneMANIA genes APOE and *LDLRAP1* are shown in the same biological pathway, while *PICALM* has the same expression patterns related to *APOE*. Of these, HIP1 and LDL-RAP1 are two examples of genes ranked by RF in both MCI-100 and AD-100 studies, which were not previously catalogued in AlzGene or passed the Fisher Exact test. More details of the relation of lipids and Alzheimer can be found in the following reviews [22] [21] [23].

5.2   Endocytosis Mediated by Clathrin

Normal cells include many mechanisms for regulation of endocytosis. Evidence indicates an important role for clathrin-mediated endocytosis in normal neurons. Abnormalities in its operation can cause neurological disorders. Specifically, clathrin plays a central role in the formation of β-amyloid. The formation of senile plaques is a common feature in the pathology and cerebral aging. In addition, the components of neurofibrillary degeneration advances the progression of AD. β-amyloid derived from APP is responsible for the formation of complexes plaques in the brain. The presence of polymorphisms in genes associated with clathrin-mediated endocytosis has been reported in patients with bipolar disorder, schizophrenia and AD, but still not enough is known to establish the process of neurodegeneration. Between the genes identified in this process, we can highlight the *HIP1, PICALM, AMPH, SH3GL2, SNAP91, BIN1, SYNJ1, SYNJ2, CBL* and *LDLRAP1*. The genes, *PICALM, AMPH* and *BIN1* are directly related to the process of endocytosis mediated by clathrin. Although there is evidence of *SH3GL2* gene as an active component in the progress of AD. Increased expression levels of *SH3GL2* in neural regions is associated with increased activation of the stress kinase c-Jun N-terminal kinase, with subsequent death of neurons. GeneMania used expression evidence to indicate that endophilin I interacts genetically with β-amyloid and *ABAD*. The AD-100 RF ranking lists SNPs close to genes HIP1,PICALM, AMPH,SH3GL2,SYNJ2 and CBL, which were not previously annotated in AlzGene or passed the Fisher Exact test. The MCI-100 RF ranking lists SNPs close to HIP1, PICALM, CBL and LDLRAB1, which were also not annotated to AlzGene or passed the Fisher Exact test. Further details of this synthesis may be found in greater detail in the following reviews [24] [25] [23].

6 Conclusions

The method RF was applied to several datasets of SNPs to check performance prediction of phenotype of patients. The results for classification tasks have shown that performance of SNPs not associated with AD, enables lower rates of mistakes in predictions. This indicates that the potential variants may contribute to the progression of AD and MCI and validates the approach for using gene networks to filter SNPs previous to RF. Additionally, the application of a filter by LD could further improve the prediction error. In comparison to a standard univariate method, the ranking of SNPs from RF indicated several novel SNPs, which were not previously catalogued in AlzGene nor passed the Fisher exact tests. Of particular interest were SNPs related to 2 genes associated to both lipids/endocytosis and 4 genes related to endocytosis previously not related to Alzheimer. These SNPs are interesting potential novel markers for the Alzheimer disease.

As a future work we would like to further explore the association of the SNPs indicated by Random Forest, such as the characterization of SNPs with main, marginal and epistatic effects and perform a biological validation of the novel candidates. From a methodological point of view, we would like to replicate the study with independent data sets and further explore the effects of model parameters on the overall ranking performance. Another interesting methodological improvement would be the use of machine learning methods that would used the network information explicitly during their feature selection steps.

Acknowledgements. Authors would like to thank CNPq, FACEPE and the Interdiciplinary Center for Clinical Research (IZKF), RWTH University Hospital for support. Data collection and sharing for this project was funded by the Alzheimer's Disease Neuroimaging Initiative (ADNI) (National Institutes of Health Grant U01 AG024904). ADNI is funded by the National Institute on Aging, the National Institute of Biomedical Imaging and Bioengineering, and through generous contributions from the following: Abbott; Alzheimers Association; Alzheimers Drug Discovery Foundation; Amorfix Life Sciences Ltd.; AstraZeneca; Bayer HealthCare; BioClinica, Inc.; Biogen Idec Inc.; Bristol-Myers Squibb Company; Eisai Inc.; Elan Pharmaceuticals Inc.; Eli Lilly and Company; F. Hoffmann-La Roche Ltd and its affiliated company Genentech, Inc.; GE Healthcare; Innogenetics, N.V.; IXICO Ltd.; Janssen Alzheimer Immunotherapy Research & Development, LLC.; Johnson & Johnson Pharmaceutical Research & Development LLC.; Medpace, Inc.; Merck & Co., Inc.; Meso Scale Diagnostics, LLC.; Novartis Pharmaceuticals Corporation; Pfizer Inc.; Servier; Synarc Inc.; and Takeda Pharmaceutical Company. The Canadian Institutes of Health Research is providing funds to support ADNI clinical sites in Canada. Private sector contributions are facilitated by the Foundation for the National Institutes of Health (www.fnih.org). The grantee organization is the Northern California Institute for Research and Education, and the study is coordinated by the Alzheimer's Disease Cooperative Study at the University of California, San Diego. ADNI data are disseminated by the Laboratory for Neuro Imaging at the University of California, Los Angeles. This research was also supported by NIH grants P30 AG010129 and K01 AG030514.

References

1. Thies, W., Bleiler, L.: Alzheimers disease facts and figures. Alzheimer's & Dementia: The Journal of the Alzheimer's Association 7, 208–244 (2011)
2. Wang, W.Y.S., Barratt, B.J., Clayton, D.G., Todd, J.A.: Genome-wide association studies: theoretical and practical concerns. Nature Reviews. Genetics 6, 109–118 (2005)
3. Bertram, L., McQueen, M.B., Mullin, K., Blacker, D., Tanzi, R.E.: Systematic meta-analyses of Alzheimer disease genetic association studies: the AlzGene database. Nature Genetics 39, 17–23 (2007)
4. Saykin, A.J., et al.: Alzheimer's Disease Neuroimaging Initiative biomarkers as quantitative phenotypes: Genetics core aims, progress, and plans. Alzheimer's & Dementia: The Journal of the Alzheimer's Association 6, 265–273 (2010)
5. Petersen, R.C., et al.: Alzheimer's Disease Neuroimaging Initiative (ADNI) Clinical characterization. Neurology 74, 201–209 (2010)
6. Kim, S., Misra, A.: SNP genotyping: technologies and biomedical applications. Annual Review of Biomedical Engineering 9, 289–320 (2007)
7. Montojo, J., Zuberi, K., Rodriguez, H., Kazi, F., Wright, G., Donaldson, S.L., Morris, Q., Bader, G.D.: GeneMANIA Cytoscape plugin: fast gene function predictions on the desktop. Bioinformatics 26(22), 2927–2928 (2010)
8. Ritchie, M.D.: Using biological knowledge to uncover the mystery in the search for epistasis in genome-wide association studies. Ann. Hum. Genet. 75(1), 172–182 (2011)
9. Mostafavi, S., Ray, D., Warde-Farley, D., Grouios, C., Morris, Q.: GeneMANIA: a real-time multiple association network integration algorithm for predicting gene function. Genome. Biol. 9(suppl. 1), S4 (2008)
10. Goldstein, B.A., Hubbard, A.E., Cutler, A., Barcellos, L.F.: An application of Random Forests to a genome-wide association dataset: methodological considerations & new findings. BMC Genetics 11, 49 (2010)
11. Lunetta, K.L., Hayward, L.B., Segal, J., Van Eerdewegh, P.: Screening large-scale association study data: exploiting interactions using random forests. BMC Genet. 5, 32 (2004)
12. Meng, Y.A., Yu, Y., Cupples, L.A., Farrer, L.A., Lunetta, K.L.: Performance of random forest when SNPs are in linkage disequilibrium. BMC Bioinformatics 10, 78 (2009)
13. Purcell, S., Neale, B., Todd-Brown, K., et al.: PLINK: a tool set for whole-genome association and population-based linkage analyses. Am. J. Hum. Genet. 81, 559–575 (2007)
14. Liaw, A., Wiener, M.: Classification and Regression by randomForest. R News 2, 18–22 (2002)
15. Breiman, L.: Random Forests. Machine Learning 45, 5–32 (2001)
16. Heidema, A.G., Boer, J.M., Nagelkerke, N., Mariman, E.C., van der A, D.L., Feskens, E.J.: The challenge for genetic epidemiologists: how to analyze large numbers of SNPs in relation to complex diseases. BMC Genet. 7, 23 (2006)
17. Glaser, B., Nikolov, I., Chubb, D., Hamshere, M.L., Segurado, R., Moskvina, V., Holmans, P.: Analyses of single marker and pairwise effects of candidate loci for rheumatoid arthritis using logistic regression and random forests. BMC Proc. 1(suppl. 1), S54 (2007)
18. Liu, C., Ackerman, H.H., Carulli, J.P.: A genome-wide screen of gene-gene interactions for rheumatoid arthritis susceptibility. Hum. Genet. 129(5), 473–485 (2011)

19. Sun, Y.V., Cai, Z., Desai, K., Lawrance, R., Leff, R., Jawaid, A., Kardia, S.L., Yang, H.: Classification of rheumatoid arthritis status with candidate gene and genome-wide single-nucleotide polymorphisms using random forests. BMC Proc. 1(suppl. 1), S62 (2007)
20. Araujo, G., Costa, I.G., Souza, M., Oliveira, J.R.M.: An Experimental Application of Random Forest on ADNI Genotype Dataset. In: Digital Proceedings of Brazilian Symposium on Bioinformatics, Campo Grande, pp. 68–73. SBC, Porto Alegre (2012)
21. Di Paolo, G., Kim, T.W.: Linking lipids to Alzheimer's disease: cholesterol and beyond. Nat. Rev. Neurosci. 12(5), 284–296 (2011)
22. Hirsch-Reinshagen, V., Burgess, B., Wellington, C.: Why lipids are important for Alzheimer disease? Molecular and Cellular Biochemistry 326(1), 121–129 (2009)
23. Holtzman, D.M., Herz, J., Bu, G.: Apolipoprotein e and apolipoprotein e receptors: normal biology and roles in Alzheimer disease. Cold Spring Harb. Perspect. Med. 2(3), a006312(2012)
24. Wu, F., Yao, P.J.: Clathrin-mediated endocytosis and Alzheimer's disease: an update. Ageing Res. Rev. 8(3), 147–149 (2009)
25. McMahon, H.T., Boucrot, E.: Molecular mechanism and physiological functions of clathrin-mediated endocytosis. Nat. Rev. Mol. Cell Biol. 12(8), 517–533 (2011)
26. Chatr-Aryamontri, A., Breitkreutz, B.J., Heinicke, S., Boucher, L., Winter, A., Stark, C., Nixon, J., Ramage, L., Kolas, N., O'Donnell, L., Reguly, T., Breitkreutz, A., Sellam, A., Chen, D., Chang, C., Rust, J., Livstone, M., Oughtred, R., Dolinski, K., Tyers, M.: The BioGRID interaction database: 2013 update. Nucleic Acids Res. 41(Database issue), D816-D823 (2013)
27. Barrett, T., Wilhite, S.E., Ledoux, P., Evangelista, C., Kim, I.F., Tomashevsky, M., Marshall, K.A., Phillippy, K.H., Sherman, P.M., Holko, M., Yefanov, A., Lee, H., Zhang, N., Robertson, C.L., Serova, N., Davis, S., Soboleva, A.: NCBI GEO: archive for functional genomics data sets–update. Nucleic Acids Res. 41(Database issue), D991-D995 (2013)
28. Cerami, E.G., Gross, B.E., Demir, E., Rodchenkov, I., Babur, O., Anwar, N., Schultz, N., Bader, G.D., Sander, C.: Pathway Commons, a web resource for biological pathway data. Nucleic Acids Res. 39(Database issue), D685-D690 (2011)
29. Brown, K.R., Jurisica, I.: Online Predicted Human Interaction Database. Bioinformatics 21(9), 2076–2082 (2005)
30. Bush, W.S., Moore, J.H.: Chapter 11: Genome-wide association studies. PLoS Comput. Biol. 8(12), e1002822 (2012)

Multilayer Cluster Heat Map Visualizing Biological Tensor Data

Atsushi Niida, Georg Tremmel, Seiya Imoto, and Satoru Miyano

Human Genome Center, Institute of Medical Science, University of Tokyo,
4-6-1 Shirokanedai, Minato-ku, Tokyo 108-8639, Japan

Abstract. Recent advances in biological high-throughput technology
are generating a broad range of omics data. Facing a torrent of massive biological data, visual data mining can be considered an intuitive
and powerful approach for hypothesis generation. The cluster heat map
approach has been popularly used to visualize the matrix types of biological data. In this study, we extended the use of the cluster heat map
to reveal informative patterns hidden in third-order tensor-type biological data. By applying the extended method, a multilayer cluster heat
map, to trans-omics and network tensor data, we successfully demonstrated the proof-of-concept of our approach. Our new visual data mining
method will be a useful tool for increasing the amount of biological tensor data. Our implementation and the tensor data studied are available
from http://www.hgc.jp/~niiyan/MCHM.

1 Introduction

Hypothesis generation from genome-wide data is currently a critical step in biological researches, for which visualization approaches are widely used. For example, the cluster heat map has been commonly used for analysis of microarray
data. While the cluster heat map has a long history itself [1], Eisen *et al.* [2]
rediscovered it in the biological field for microarray data analysis. In a typical
microarray experimental design, genome-wide gene expression across multiple
samples are profiled to produce a matrix whose rows and columns represent
genes and samples. The cluster heat map reorders the rows and columns of the
matrix based on hierarchical clustering, and visualizes values using color codes.

Recently, new technologies represented as next-generation sequencing have en-
abled comprehensive profiling of the various types of cellular variables assigned
to each gene: mRNA quantity, protein quantity, TF binding, histone modifications, etc. Furthermore, several important project such as TCGA and ENCODE
have been launched to systematically perform trans-omics profiling across dozens
to hundreds of samples [3, 4]. We obtain *gene × sample* matrices for all targeted
cellular variables as outputs from the trans-omics profiling projects. We also infer gene regulatory networks from omics data. If we obtain multiple networks
from multiple data sources, the data are represented as multiple *gene × gene*
adjacency matrices [5]. Both the trans-omics and network data are represented

J.C. Setubal and N.F. Almeida (Eds.): BSB 2013, LNBI 8213, pp. 116–125, 2013.

as multiple matrices sharing common row and column elements. By stacking the matrices, we can also regard the data as third-order tensors.

A tensor is a multidimensional array. While first and second-order tensors are a vector and matrix, a third-order tensor has 3 indices, each of which is on one of the 3 dimensions called the first, second and third orders. Integrative analysis of tensor data is expected to produce more rich hypotheses than separate analysis of each individual component matrix. However, only a few methods has been designed for analysis of tensor-type biological data. Moreover, because most of them are largely based on the mathematical approach, it is often difficult to interpret obtained results and gain biological insights from them [6, 7].

In this paper, we propose a new visual data mining method, the multilayer cluster heat map, to analyse tensor-type biological data. The multilayer cluster heat map is an extension of the ordinary cluster heat map for matrix data, used to to analyze third-order tensor data. Application of the multilayer cluster heat map to TCGA, ENCODE, and cancer co-expression network data demonstrated that our visual data mining method is useful for generating biological hypotheses from biological tensor data.

2   Materials and Methods

2.1   Multilayer Cluster Heat Map

Assume that we perform experimental profiling of l cellular variables assigned to n genes across m samples. We then obtain an $n \times m \times l$ *trans-omics tensor* $\mathbf{T}$, where element $\mathbf{T}(i, j, k)$ is the value of variable k for gene i in sample j. When we have multiple network data, each of which is represented by an adjacency matrix, the data are represented as the same types of tensor data. However, in the *network tensor*, 2 of the 3 order have an equal dimension, which is the number of genes in the networks. Namely, if you have l networks containing n genes, the data are an $n \times m \times l$ tensor with $m = n$, where element $\mathbf{T}(i, j, k)$ represents the edge between gene i and gene j in network k.

We define a *slice* matrix of the tensor as all elements that have the same index at an order. Namely, if a colon specifies the set of all indices for a particular order, $\mathbf{T}(i, :, :)$, $\mathbf{T}(:, j, :)$, and $\mathbf{T}(:, :, k)$ are slice matrices at the first, second and third order, respectively. In this study, we also refer to each slice using its associated dimension of the data: e.g., gene-order slices, sample-order slices, etc.

Prior to visualization, the multilayer cluster heat map reorders slices along each order based on hierarchical clustering. We calculate the squared Euclidian distance between 2 first order slices, which correspond to genes i_1 and i_2 in trans-omics or network tensor data, as follows:

$$d^2(\mathbf{T}(i_1, :, :), \mathbf{T}(i_2, :, :)) = \sum_{j=1}^{m} \sum_{k=1}^{l} (\mathbf{T}(i_1, j, k) - \mathbf{T}(i_2, j, k))^2$$

We perform hierarchical clustering of the first-order slices based on Ward's method employing this distance measure. The first-order slices are reordered according to the clustering result. The second and third-order slices are reordered similarly. The slices of each order are visualized using color codes. We implemented interactive application for visualization of the heat maps using JAVA and the Processing library (http://processing.org/), and which is available from http://www.hgc.jp/~niiyan/MCHM.

2.2   Data Preparation

TCGA Data. For glioblastoma primary tumor samples, we downloaded mRNA expression, DNA methylation, and DNA copy number data from the TCGA data portal (https://tcga-data.nci.nih.gov/tcga/) [4]. For mRNA expression and DNA methylation data, BI HT_HG-U133A and JHU-USC HumanMethylation27 were obtained. After the values were transformed to the logarithmic scale and normalized so that the mean is 0 and the variance 1 in each sample, the probe set IDs were converted to gene symbols. For DNA copy number data, BI Genome_Wide_SNP_6 was obtained. The segmented copy number values across chromosomes were converted to those for each gene, and the values were normalized so that the elements of the *gene* × *sample* copy number matrix have mean 0 and variance 1. For each of the genes existing in the three types of omics data, three types of omics values across all samples were obtained, and their variances were calculated. For top 500 genes with the largest variances, the three types of omics matrices were extracted and renormalized so that the mean was 0 and the variance 1 in each matrix. The processed matrices were combined as a *gene* × *sample* × *variable* tensor and subjected to the multilayer cluster heat map.

ENCODE Data. The histone modification profiles of 7 cell lines were downloaded from the Broad Histone Track at the UCSC ENCODE site (http://genome.ucsc.edu/ENCODE/) [3]. After broadPeak files were obtained for 8 types of histone modification profiles across the 7 cell lines, each peak was associated with gene promoter regions. We associated a peak with a gene if the peak overlapped the promoter region of the gene, which we assumed covered ±500 bp around its transcription start site. We then obtained a binary third order tensor containing the histone modification profiles of genes across the 7 cell lines. From this *gene* × *sample* × *variable* binary tensor, we extracted a sub-tensor for the top 3000 variable gene-order slices in the same way for the TCGA data and visualized it.

Cancer Co-expression Network Data. We obtained 260 mRNA expression data sets from the GEO database [8]. They are associated with the keyword "cancer", published after 2005, obtained by major microarray platforms, and contains ≥ 50 samples. First, to obtain a weighted co-expression network for each dataset, we calculated Pearson's correlation coefficients c between every gene pair as edge weights. They were then subjected to Fisher's transformation, which is calculated as $z_{ij} = \frac{1}{2}\log\frac{1+c_{ij}}{1-c_{ij}}$, where c_{ij} is the correlation between genes i and j. We also calculated the average of absolute edge weights as the hubness for each gene; i.e., $h_i = \sum_{j \in \{1,\cdots,i-1,i+1,\cdots,n\}} |z_{ij}|/(n-1)$, where n is the number of genes. The hubness for each gene was summed across the networks, and the top 1000 genes with the largest hubness sum were selected. Similarly, the hubnesses of all genes was summed in each of the networks and the top 100 networks with the largest hubness sum were selected. For the 1000 genes, adjacency matrices representing each of the 100 networks were obtained after edge weights were normalized so that the mean was 0 and the variance was 1 in each co-expression network. If a gene was missing from a dataset, edge weights associated with the genes were set to 0 in the corresponding matrix. We also set the maximum and minimum of the edge weights to 5 and -5, respectively. Finally, the processed adjacency matrices were combined as a *gene* × *gene* × *network* tensor for visualization.

3   Results

3.1   TCGA Data

First, we visualized trans-omics tensor data from TCGA. The data included mRNA expression, DNA methylation, and DNA copy number profiles for thousands of genes across approximately 200 glioblastoma clinical samples. For visualization, we focused on 500 genes with the highest variance in the data. Figure 1 illustrates 3 variable-order slices in the multilayer cluster heat map, which reveal correlation structures hidden in the omics tensor data. For example, comparison of mRNA expression and DNA copy number slices revealed several *gene* × *sample* biclusters with high values for both mRNA expression and DNA copy number, suggesting that DNA copy number amplification led to an up-regulation of mRNA expression in some sample groups. Similarly, we found a large *gene* × *sample* bicluster where high promoter methylation was associated with repression of mRNA expression. In glioblastoma, it is known that a sample subgroup is associated with the CpG island hyper-methylation phenotype [4]. The heat maps revealed such a sample subgroup and genes that were both methylated and transcriptionally repressed in the sample subgroup. Collectively, these results demonstrate the usefulness of the multilayer cluster heat map to reveal correlation structures among multiple omics layers at a glance.

3.2   ENCODE Data

Next, we applied the multilayer cluster heat map to the ENCODE trans-omics tensor data. The data contain 7 histone modification profiles for 3000 genes across 7 cell lines, described in Table 1. The tensor data contain binary values indicating whether the promoter of each gene has each type of modification. Figure 2 depicts sample-order slices of the tensor and the dendrograms for the 3-way clusterings. The sample-order clustering divides the samples, reflecting their cellular origin: e.g., the 2 blood-derived cell lines, K562 and Gm12878, form a distinct cluster.

The histone modification types were also clustered according to their functionality. It is known that methylation of H3k4 and acetylation of H3k27 and H3k9 are

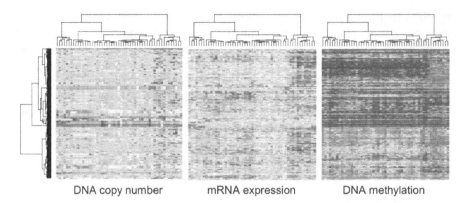

DNA copy number   mRNA expression   DNA methylation

Fig. 1. Visualization of the TCGA data. A third-order tensor containing DNA copy number, mRNA expression, and DNA methylation was prepared from TCGA data. The variable-order slices of the trans-omics tensor are displayed as heat maps, where red and blue represent high and low values, respectively. Rows and columns of the sliced matrices represent genes and samples, respectively.

associated with transcriptional activation. These modifications form an activation-associated major cluster, that contains 2 sub-clusters: the H3k4 methylation cluster (H3k4me1, H3k4me2 and H3k4me3) and the H3 acetylation cluster (H3k27ac and H3k9ac). The sample-order slices showed that the modification profiles for the 2 sub-clusters possessed different degrees of similarity across the 7 cell lines. For example, the blood-derived cell lines possessed similar profiles between the 2 sub-cluster, while the ectoderm-derived cell lines (Nhek and Hmec) formed a gene cluster that was modified at the H3k4 methylation cluster, but not at the H3 acetylation cluster. H3k27me3 is known to be associated with transcriptional repression, and its profiles are inversely correlated with those of the activation-associated major cluster. H3k36me3 and H4k20me also exhibited unique profiles. H3k36me3 shows relatively constant profiles across cell lines. On the other hand, overall degrees of H4k20me1 modification differed across cell lines: K562 has many target genes of H4k20me1, while Nhlf has only a few.

Based on these histone modification patterns across cell lines, the genes were divided into 5 clusters. The gene clusters are represented by color bars in Figure 2, and their information is provided in Table 2. We also examined the enrichment of GO terms in each cluster using the hypergeometric test. We could not identify clear GO enrichment in all clusters, possibly because the number of cell lines in the ENCODE data set was small. However, one gene cluster (cluster 4) was clearly enriched for the GO term "ectoderm development". Interestingly, the gene cluster overlapped the aforementioned gene cluster that possessed characteristic profiles for the H3k4 methylation and H3 acetylation cluster in ectoderm cell lines. Taken together, these results demonstrate that the multilayer cluster heat map enables integrative interpretation of trans-omics profiles across multiple samples, which could not be achieved by conventional approaches targeting matrix-type data.

Table 1. Cell lines in the ENCODE data

Name	Description	Lineage	Tissue
Nhek	epidermal keratinocytes	ectoderm	skin
Hmec	mammary epithelial cells	ectoderm	breast
Hsmm	skeletal muscle myoblasts	mesoderm	muscle
Nhlf	lung fibroblasts	endoderm	lung
Huvec	umbilical vein endothelial cells	mesoderm	blood vessel
Gm12878	B-lymphocyte, lymphoblastoid	mesoderm	blood
K562	chronic myelogenous leukemia	mesoderm	blood

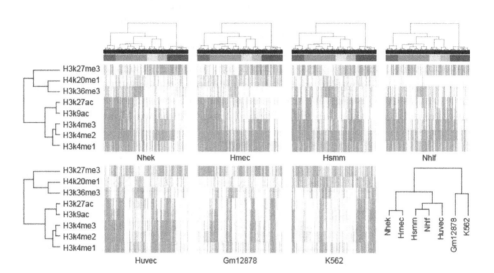

Fig. 2. Visualization of the ENCODE data. A third-order tensor containing 7 histone modification profiles across 7 cell lines was prepared from ENCODE data. The sample-order slices of the trans-omics tensor are displayed as heat maps with dendrograms from 3-way clusterings. In the heat maps, red and white represent the presence and absence of the modifications, respectively. Rows and columns of the sliced matrices represent modification types and genes, respectively. A color band below the gene-order dendrogram indicates gene clusters whose information is provided in Table 2.

Table 2. Gene clusters in the ENCODE data

ID	Size	Color	Most significantly enriched GO term	p-value
cluster 1	433	red	immune system process	4.15×10^{-4}
cluster 2	1115	orange	collagen	1.19×10^{-5}
cluster 3	382	yellow		
cluster 4	344	green	ectoderm development	1.19×10^{-10}
cluster 5	726	blue		

3.3   Cancer Co-expression Network Data

Lastly, we attempted to visualize cancer co-expression network data using the multilayer cluster heat map. We obtained co-expression networks for 260 microarray data sets, and combined adjacency matrices representing each network as a third-order tensor. Figure 3 illustrates 10 representatives of the network-order slices and dendrograms for the network and gene-order clusterings. Supplementary Figure 1 presents an overview of all 100 network-order slices. The network-order clustering divided the networks into 2 major clusters. The members of one network cluster exhibited clear gene clusters while those of the other show less clear gene clusters. The gene-order clustering revealed recurrent module structures in the cancer co-expression network. Information regarding the 4 gene modules found is provided in Table 3. The most recurrent module (cluster 1) was highly enriched for the GO term "cell cycle" while the second recurrent module (cluster 4) was enriched for "immune system process". The network-order slices revealed which network harbored these modules in a glance. These results demonstrate the usefulness of the multilayer cluster heat map in analyzing not only trans-omics data but also multiple network data.

Table 3. Gene clusters in the cancer co-expression network data

ID	Size	Color	Most significantly enriched GO term	p-value
cluster 1	67	red	cell cycle	1.09×10^{-18}
cluster 2	266	orange	cytoskeletal protein binding	4.46×10^{-6}
cluster 3	272	green	apoptotic mitochondria changesl	1.52×10^{-4}
cluster 4	393	blue	immune system process	6.28×10^{-8}

4   Discussion

In this study, we present the multilayer cluster heat map, a novel visual data mining method for biologists working with biological tensor data. We applied the multilayer cluster heat map to trans-omics and network tensor data, and successfully demonstrated the proof-of-concept of our approach. However, a number of issue must be to be addressed. For example, it should be noted that clustering results are highly dependent on the method of data preprocessing. In particular, when dealing with trans-omics data the fact that each type of omics data have different scales and distributions of measurements should be taken into consideration. Moreover, results depend on the distance measure and clustering method, as is the case in conventional cluster heat maps. Thus, it is desirable to establish some criteria to select optimal preprocessing and clustering methods for each type of data. Otherwise, different preprocessing and clustering methods should be attempted to assure the robustness of obtained conclusions. In this study, we confirmed that even when different preprocessing and clustering methods are employed, principal findings are essentially unchanged in spite of different heat map appearances.

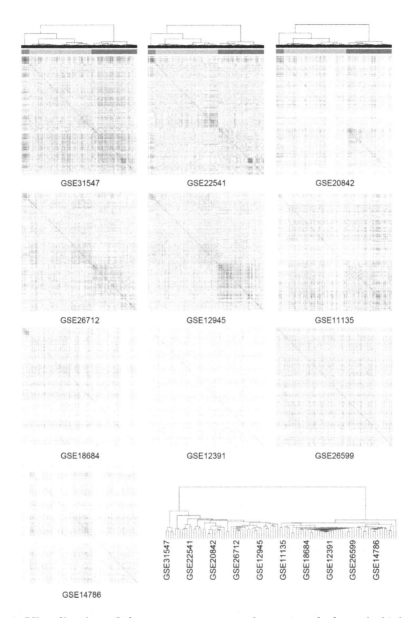

Fig. 3. Visualization of the cancer co-expression network data. A third-order tensor containing 100 cancer co-expression networks was prepared from the GEO database. The 10 representative network-order slices of the network tensor are displayed with gene-order and network-order dendrograms. In the heat maps, red and blue represent the positive and negative correlations. Rows and columns of the sliced matrices represent genes, A color band below the gene-order dendrogram indicates gene clusters whose information is contained in Table 3.

Supplementary Figure 1. Visualization of the cancer co-expression network data. All 100 network-order slices of the cancer co-expression network are arranged according to the network-order dendrogram in Figure 3 and displayed left-to-right and top-to-bottom.

Another challenge is the computational implementation. Since biological tensor data containing the gene-order is intrinsically large, their cluster heat map visualization is a computationally intensive problem. As our current implementation does not provide sufficient efficiency to visualize the whole space of the input data, we focus only on a small sub-set of the data space by filtering out less informative spaces in the preprocessing steps. However, since such stringent filtering hides much information, it is more preferable to visualize a larger data space by employing less stringent filtering and more efficient implementation.

The visualization approach is powerful since it enables biologists without mathematical knowledge to intuitively interpret complex information hidden in massive amounts of data. On the other hand, the mathematical approach is also a necessary complement. It can systematically extract hidden information and quantitatively evaluate the result. By combining these 2 complementary approaches, we will obtain more fruitful results, as demonstrated by the history of microarray analyses, which employs heat map visualization combined with various statistical methods. We expect that our multilayer cluster heat map can provide more insights when combined with tensor mathematics [9].

While third-order tensor types of data are becoming more common in the field of biology, tensor data analysis is still considered as a challenge. We believe that our visual data mining method will be useful for obtaining biological insights from massive amounts of biological tensor data.

Acknowledgments. Computation time was provided by the Super Computer System, Human Genome Center, Institute of Medical Science, University of Tokyo. This work was partially supported by "Systems Cancer" (Project No 4201), a Grant-in-Aid for Scientific Research on Innovative Areas, and the Global COE Program "Center of Education and Research for the Advanced Genome-Based Medicine - For personalized medicine and the control of worldwide infectious diseases -" by MEXT, Japan.

References

[1] Wilkinson, L., Friendly, L.: The history of the cluster heat map. The American Statistician 63, 179–184 (2009)
[2] Eisen, M., Spellman, P., Brown, P., Botstein, D.: Cluster analysis and display of genome-wide expression patterns. Proc. Natl. Acad. Sci. U. S. A. 95, 14863–14868 (1998)
[3] Rosenbloom, K., Dreszer, T., Long, J., Malladi, V., Sloan, C., Raney, B., Cline, M., Karolchik, D., Barber, G., Clawson, H., Diekhans, M., Fujita, P., Goldman, M., Gravell, R., Harte, R., Hinrichs, A., Kirkup, V., Kuhn, R., Learned, K., Maddren, M., Meyer, L., Pohl, A., Rhead, B., Wong, M., Zweig, A., Haussler, D., Kent, W.: ENCODE whole-genome data in the UCSC Genome Browser: update 2012. Nucleic Acids Res. 40, D912–D917 (2012)
[4] Noushmehr, H., Weisenberger, D., Diefes, K., Phillips, H., Pujara, K., Berman, B., Pan, F., Pelloski, C., Sulman, E., Bhat, K., Verhaak, R., Hoadley, K., Hayes, D., Perou, C., Schmidt, H., Ding, L., Wilson, R., Van Den Berg, D., Shen, H., Bengtsson, H., Neuvial, P., Cope, L., Buckley, J., Herman, J., Baylin, S., Laird, P., Aldape, K.: Identification of a CpG island methylator phenotype that defines a distinct subgroup of glioma. Cancer Cell 17, 510–522 (2010)
[5] Li, W., Liu, C., Zhang, T., Li, H., Waterman, M., Zhou, X.: Integrative analysis of many weighted co-expression networks using tensor computation. PLoS Comput. Biol. 7, e1001106 (2011)
[6] Omberg, L., Golub, G., Alter, O.: A tensor higher-order singular value decomposition for integrative analysis of DNA microarray data from different studies. Proc. Natl. Acad. Sci. U. S. A. 104, 18371–18376 (2007)
[7] Alter, O., Golub, G.: Reconstructing the pathways of a cellular system from genome-scale signals by using matrix and tensor computations. Proc. Natl. Acad. Sci. U. S. A. 102, 17559–17564 (2005)
[8] Barrett, T., Troup, D., Wilhite, S., Ledoux, P., Evangelista, C., Kim, I., Tomashevsky, M., Marshall, K., Phillippy, K., Sherman, P., Muertter, R., Holko, M., Ayanbule, O., Yefanov, A., Soboleva, A.: NCBI GEO: archive for functional genomics data sets–10 years on. Nucleic Acids Res. 10, D1005–D1010 (2005)
[9] Tamara, G., Brett, W.: Tensor decompositions and applications. SIAM Review 51, 455–500 (2009)

On the 1.375-Approximation Algorithm
for Sorting by Transpositions in $O(n \log n)$ Time

Luís Felipe I. Cunha[1], Luis Antonio B. Kowada[2],
Rodrigo de A. Hausen[3], and Celina M.H. de Figueiredo[1]

[1] Universidade Federal do Rio de Janeiro
{lfignacio,celina}@cos.ufrj.br
[2] Universidade Federal Fluminense
luis@vm.uff.br
[3] Universidade Federal do ABC
hausen@compscinet.org

Abstract. Sorting by Transpositions is an NP-hard problem. Elias and Hartman proposed a 1.375-approximation algorithm, the best ratio so far, which runs in $O(n^2)$ time. Firoz *et al.* proposed an improvement to the running time, from $O(n^2)$ down to $O(n \log n)$, using Feng and Zhu's permutation trees. We provide counter-examples to the correctness of Firoz *et al.*'s strategy, showing that it is not possible to reach a component by sufficient extensions using the method proposed by them.

Keywords: comparative genomics, genome rearrangement, sorting by transpositions, approximation algorithms.

1 Introduction

By comparing the orders of common genes between two organisms, one may estimate the series of mutations that occurred in the underlying evolutionary process. In the simplified genome rearrangement model adopted in this paper, each mutation is a transposition, and the sole chromosome of each organism is modeled by a permutation, which means that there are no duplicated or deleted genes. A transposition is a rearrangement of the gene order within a chromosome, in which two contiguous blocks are swapped. A biological explanation for this rearrangement is the duplication of a block of genes, followed by the deletion of the original block [15]. The transposition distance is the minimum number of transpositions required to transform one chromosome into the other. Bulteau *et al.* [2] proved that Sorting by Transpositions (SBT), the problem of determining the transposition distance between two permutations, is NP-hard.

Our approach towards the SBT problem has been as follows: to determine the maximum value of the transposition distance among all permutations, i.e., the transposition diameter [3,4,13]; to obtain tight bounds for the transposition distance of some classes of permutations [3,11,13]; and to explore approximation algorithms for estimating the transposition distance between permutations in general, providing better practical results or lowering their time complexities [14].

J.C. Setubal and N.F. Almeida (Eds.): BSB 2013, LNBI 8213, pp. 126–135, 2013.

Two important measures are important for the study of approximation algorithms: the approximation ratio and the running time of such algorithms. In the following paragraphs, we discuss the existing approximation algorithms for the SBT problem and their respective measures.

Bafna and Pevzner [1] designed the first 1.5-approximation algorithm with an $O(n^2)$ running time, based on the cycle structure of the *breakpoint graph*. Hartman and Shamir [10], by introducing simple permutations, proposed an easier 1.5-approximation algorithm and, by exploiting a balanced tree data structure designed by Kaplan and Verbin [12], decreased the running time to $O(n^{\frac{3}{2}} \sqrt{\log n})$. Feng and Zhu [7] developed the balanced *permutation tree*, further decreasing the complexity of Hartman and Shamir's 1.5-approximation to $O(n \log n)$, and Lopes *et al.* [14] implemented this algorithm.

Elias and Hartman [6] obtained, by a thorough computational case analysis of the cycles of the breakpoint graph, an 1.375-approximation $O(n^2)$ algorithm, which was implemented by Dias and Dias [5]. Firoz *et al.* [8] recently claimed that this 1.375-approximation algorithm can run in $O(n \log n)$ time using the permutation tree data structure developed by Feng and Zhu.

In the present paper, we show that Firoz *et al.*'s usage of a query procedure to extend a full configuration to a component fails in some situations. We provide an infinite family for which Firoz *et al.*'s algorithm does not find a $\frac{11}{8}$-sequence, proving that their strategy for running the 1.375-approximation algorithm in $O(n \log n)$ time is incorrect.

This article is organized as follows: in Section 2 we present the breakpoint graph, its relevant properties and how it can be used to sort a permutation by transpositions; Section 2.2 is devoted to Elias and Hartman's 1.375-approximation algorithm; Section 2.3 presents the permutation tree data structure; Section 3 describes Firoz *et al.*'s approach for using the permutation tree in the 1.375-approximation algorithm and counterexamples to the correctness of their approach; Section 4 contains the final remarks of this paper.

2 Background

For our purposes, a gene is represented by a unique integer and a chromosome with n genes is a permutation $\pi = [\pi_0\ \pi_1\ \pi_2\ \cdots\ \pi_n\ \pi_{n+1}]$, where $\pi_0 = 0$, $\pi_{n+1} = n+1$ and each π_i is a unique integer in the range $1 \ldots n$. The *transposition* $t(i, j, k)$, where $1 \leq i < j < k \leq n+1$, is the permutation $[0\ 1 \ldots i{-}1\ j\ j{+}1 \ldots k{-}1\ i\ i{+}1 \ldots j{-}1\ k\ k{+}1 \ldots n\ n{+}1]$. The *application* of a transposition $t(i, j, k)$ to a permutation π is the product $\pi\, t(i, j, k)$, denoted as an action to the right; it interchanges the two contiguous blocks $\pi_i\ \pi_{i+1} \ldots \pi_{j-1}$ and $\pi_j\ \pi_{j+1} \ldots \pi_{k-1}$. A sequence of transpositions $t(i_1, j_1, k_1), t(i_2, j_2, k_2), \ldots, t(i_q, j_q, k_q)$ *sorts* a permutation π if $\pi\, t(i_1, j_1, k_1)\, t(i_2, j_2, k_2) \cdots t(i_q, j_q, k_q) = \iota$, where ι is the identity permutation $[0\ 1\ 2 \ldots n\ n+1]$. The *transposition distance* of π, denoted $d(\pi)$, is the length of a minimum sequence of transpositions that sorts π.

2.1   The Breakpoint Graph

Given a permutation π, its *breakpoint graph* is the graph $G(\pi) = (V, R \cup D)$. The set of vertices V is $\{0, -1, +1, -2, +2, \cdots, -n, +n, -(n+1)\}$, and the set of undirected edges is partitioned into two sets, the *reality edges* R and the *desire edges* D, where $R = \{b_i = (+\pi_i, -\pi_{i+1}) \mid i = 0, \ldots, n\}$ and $D = \{(+i, -(i+1)) \mid i = 0, \ldots, n\}$. Fig. 1 shows the diagram for the breakpoint graph $G([\mathbf{0}\,10\,9\,8\,7\,1\,6\,11\,5\,4\,3\,2\,\mathbf{12}])$, where the horizontal lines represent the edges in R and the arcs represent the edges in D. The permutation $[\mathbf{0}\,10\,9\,8\,7\,1\,6\,11\,5\,4\,3\,2\,\mathbf{12}]$ is a key permutation for our proposed conter-examples.

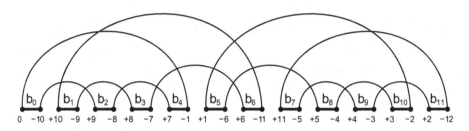

Fig. 1. $G([\mathbf{0}\ 10\ 9\ 8\ 7\ 1\ 6\ 11\ 5\ 4\ 3\ 2\ \mathbf{12}])$

As a direct consequence of the definition, every vertex in $G(\pi)$ has degree two, so $G(\pi)$ can be partitioned into disjoint cycles. We shall use the terms *a cycle in* π and *a cycle in* $G(\pi)$ interchangeably to denote the latter. We say that a cycle in π has length ℓ, or that it is an ℓ-*cycle*, if it has exactly ℓ reality edges (or, equivalently, ℓ desire edges). A permutation π is a *simple permutation* if every cycle in π has length at most 3.

The number of cycles of odd length in $G(\pi)$, denoted by $c_{odd}(\pi)$, is an important measure. A transposition t is said to be an x-*move for* π if $c_{odd}(\pi t) = c_{odd}(\pi) + x$. Bafna and Pevzner [1] observed that a transposition deletes three reality edges in $G(\pi)$ and recreates them differently, which implies that there are only -2, 0 and 2-moves; since $c_{odd}(\iota) = n + 1$, a lower bound for $d(\pi)$ is $\left\lceil \frac{(n+1) - c_{odd}(\pi)}{2} \right\rceil$, where the equality holds if, and only if, π can be sorted with 2-moves only.

Hannenhalli and Pevzner [9] proved that every permutation π can be transformed into a simple one $\hat{\pi}$ preserving the lower bound for the distance, i. e. $\left\lceil \frac{(n+1) - c_{odd}(\pi)}{2} \right\rceil = \left\lceil \frac{(m+1) - c_{odd}(\hat{\pi})}{2} \right\rceil$ where m is such that $\hat{\pi} = [\mathbf{0}\hat{\pi}_1 \ldots \hat{\pi}_m \mathbf{m+1}]$. They have also shown that a sequence that sorts $\hat{\pi}$ can be transformed into a sequence that sorts π, which implies that $d(\pi) \leq d(\hat{\pi})$. This method is commonly used in the literature, as in Hartman and Shamir's [10] and Elias and Hartman's [6] algorithms.

A transposition $t(i, j, k)$ *affects* a cycle C if it contains one of the following reality edges: b_{i+1}, or b_{j+1}, or b_{k+1}. A cycle is *oriented* if there is a 2-move that

affects it, otherwise it is *unoriented*. If there exists a 2-move that may be applied
to π, we say that π is *oriented*, otherwise π is *unoriented*.

A sequence of q transpositions is a (q, r)-*sequence* if it has r transpositions
that are 2-moves. A $\frac{q}{r}$-sequence is a (x, y)-sequence such that $x \leq q$ and $\frac{x}{y} \leq \frac{q}{r}$.
Bafna and Pevzner [1] proved that there always exists a $(3, 2)$-sequence for any
given permutation, which implies that there is an approximation scheme for the
transposition distance with ratio $\frac{3/4}{1/2} = 1.5$, and that the search of the correct
cycles to apply the $(3, 2)$-sequence runs in quadratic time. Therefore, Bafna and
Pevzner's algorithm runs in $O(n^2)$ time. Hartman and Shamir [10] obtained a
simpler method to obtain a $(3, 2)$-sequence when transforming π into a simple
permutation $\hat{\pi}$. They also improved the running time, decreasing the complexity
of the algorithm down to $O(n^{\frac{3}{2}}\sqrt{\log n})$ time.

Interactions between Cycles. A cycle in π is determined by its reality edges,
in the order that they appear, starting from the leftmost edge. The notation
$C = \langle b_{x_1} b_{x_2} \ldots b_{x_\ell} \rangle$, where $x_1 = \min\{x_1, x_2, \ldots, x_\ell\}$, characterizes an ℓ-cycle.
For instance, in Fig. 1, the cycles are: $C_1 = \langle b_0, b_2, b_4 \rangle$, $C_2 = \langle b_1, b_3, b_6 \rangle$, $C_3 = \langle b_5, b_8, b_{10} \rangle$ and $C_4 = \langle b_7, b_9, b_{11} \rangle$.

Let b_x, b_y, b_z, where $x < y < z$, be edges in C, and $b_{x'}, b_{y'}, b_{z'}$, where $x' < y' < z'$ be edges in a different cycle C'. We say that the pairs (b_x, b_y) and $(b_{x'}, b_{y'})$
intersect if these edges occur in alternating order in the breakpoint graph, i.e.
$x < x' < y < y'$ or $x' < x < y' < y$. Similarly, two triplets of reality edges
(b_x, b_y, b_z) and $(b_{x'}, b_{y'}, b_{z'})$ are *interleaving* if their edges occur in alternating
order, i.e. $x < x' < y < y' < z < z'$ or $x' < x < y' < y < z' < z$. Two
cycles C and C' intersect if there is a pair of reality edges in C that intersects
with a pair of reality edges in C', and two 3-cycles are interleaving if their
edges interleave. From Fig. 1, one can notice that the cycles $C_2 = \langle b_1, b_3, b_6 \rangle$
and $C_3 = \langle b_5, b_8, b_{10} \rangle$ are intersecting, but C_2 and C_3 are not interleaving;
the cycles $C_1 = \langle b_0, b_2, b_4 \rangle$ and $C_2 = \langle b_1, b_3, b_6 \rangle$ are interleaving, and so are
$C_3 = \langle b_5, b_8, b_{10} \rangle$ and $C_4 = \langle b_7, b_9, b_{11} \rangle$.

A *configuration* A of π is a subset of the cycles in $G(\pi)$. A configuration A is
connected if, for any two cycles C_1 and C_k in A, there are cycles $C_1, \ldots, C_{k-1} \in A$
such that, for each $i \in \{1, 2, \ldots, k-1\}$, the cycle C_i intersects or interleaves with
C_{i+1}. If the configuration A is connected and maximal, then A is a *component*.
For instance, in Fig. 1, the configuration $\{C_1, C_2, C_3, C_4\}$ is a component, but
the configuration $\{C_1, C_2, C_3\}$ is not a component. Given a configuration A, if
there exists a permutation π such that $G(\pi)$ is isomorphic to A, we say that A
and π are *equivalent*. Every component has an equivalent permutation [6].

Let C be a 3-cycle in a configuration A. An *open gate* is a pair of reality
edges of C that does not intersect any other pair of reality edges of any cycle
in A. If a configuration A has only 3-cycles with no open gates, then A is a *full
configuration*. In Fig. 1, the configuration $A = \{C_2\}$ has two open gates, whereas
$A \cup \{C_1\}$ is a full configuration. Not every full configuration has an equivalent
permutation, as can be seen in Fig. 2.

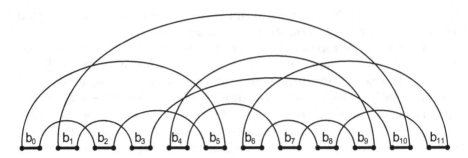

Fig. 2. A full configuration with no equivalent permutation. The cycles are $\{\langle b_0, b_2, b_5\rangle, \langle b_1, b_3, b_{10}\rangle, \langle b_4, b_7, b_9\rangle, \langle b_6, b_8, b_{11}\rangle\}$.

2.2 Elias and Hartman's 1.375-approximation Algorithm

Elias and Hartman [6] performed a systematic enumeration of all components having nine or less cycles, in which all of the cycles have length 3. The components were obtained from a single 3-cycle, after applying a series of *sufficient extensions*, as described next. An *extension* of a configuration A is a configuration $A \cup \{C\}$, where $C \notin A$ and A is connected with C, and a *sufficient extension* is an extension that either: 1) closes an open gate; or 2) extends a full configuration such that the extension has at most one open gate. Configurations obtained by series of sufficient extensions are named *sufficient configurations*.

Lemma 1. *[6] Every unoriented sufficient configuration of nine cycles has an* $\frac{11}{8}$-*sequence.*

Components with less than nine cycles are called *small components*. Elias and Hartman showed that there are just five kinds of small components that do not have an $\frac{11}{8}$-sequence; these components are called *bad small components*. Even if a permutation has bad small components, it still might be possible to find an $\frac{11}{8}$-sequence, as Lemma 2 states.

Lemma 2. *[6] Let π be a permutation with at least eight cycles that contains only bad small components. Then, π has an $(11, 8)$-sequence.*

Corollary 1. *[6] Every 3-permutation with at least eight cycles has an* $\frac{11}{8}$-*sequence.*

Lemmas 1 and 2, and Corollary 1 form the theoretical basis for Elias and Hartman's 1.375-appoximation algorithm for sorting by transpositions, described in detail in Algorithm 1.

2.3 The Permutation Tree

Feng and Zhu [7] introduced the *permutation tree*, a binary balanced tree that represents a permutation. Let $\pi = [\pi_0 \pi_1 \pi_2 \cdots \pi_n \pi_{n+1}]$ be a permutation.

Algorithm 1. Elias and Hartman's Sort(π)

---

1   Transform permutation π into a simple permutation $\hat{\pi}$.
2   Check if there is a $(2,2)$-sequence. If so, apply it.
3   While $G(\hat{\pi})$ contains a 2-cycle, apply a 2-move.
4   $\hat{\pi}$ consists of 3-cycles. Mark all 3-cycles in $G(\hat{\pi})$.
5   **while** $G(\hat{\pi})$ *contains a marked 3-cycle C* **do**
6   **if** C *is oriented* **then**
7   ⌊ Apply a 2-move to it.
8   **else**
9   Try to sufficiently extend C eight times (to obtain a configuration with at most nine cycles).
10   **if** *sufficient configuration with nine cycles has been achieved* **then**
11   ⌊ Apply an $\frac{11}{8}$-sequence. (Lemma 1)
12   **else**
13   (It is a small component.)
14   **if** *it is not a bad component* **then**
15   ⌊ Apply an $\frac{11}{8}$-sequence. (Corollary 1)
16   **else**
17   ⌊ Unmark all cycles of the component.

18 (Now $G(\hat{\pi})$ has only bad small components.)
19 **while** $G(\hat{\pi})$ *contains at least eight cycles* **do**
20   ⌊ Apply an $\frac{11}{8}$-sequence (Lemma 2)
21 While $G(\hat{\pi})$ contains a 3-cycle, apply a $(3,2)$-sequence. (Hartman and Shamir [10])
22 Mimic the sorting of π using the sorting of $\hat{\pi}$. (Hartman and Shamir [10])

The corresponding permutation tree has n leaves, labeled $\pi_1, \pi_2, \cdots, \pi_n$; every node represents an interval of consecutive elements $\pi_i, \pi_{i+1}, \cdots, \pi_{k-1}$, with $i < k$, and is labeled by the maximum number in the interval. Therefore, the root of the tree is labeled with n. Furthermore, the left child of a node represents the interval $\pi_i, \cdots, \pi_{j-1}$ and the right child represents $\pi_j, \cdots, \pi_k$, with $i < j < k$. Feng and Zhu provided algorithms: to *build* a permutation tree in $O(n)$ time, to *join* two permutation trees into one in $O(h)$ time, where h is the height difference between the trees, and to *split* a permutation tree into two in $O(\log n)$ time.

The operations *split* and *join* allow one to apply a transposition to a permutation π, updating the tree, in time $O(\log n)$. This is done as follows: i) do successive splits of T into four permutations trees T_1, T_2, T_3 and T_4, that correspond to $[\boldsymbol{\pi_0}, \pi_1, \cdots, \pi_{i-1}]$, $[\pi_i, \cdots, \pi_{j-1}]$, $[\pi_j, \cdots, \pi_{k-1}]$, and $[\pi_k, \cdots, \pi_n, \boldsymbol{\pi_{n+1}}]$, respectively; and ii) perform the joins of T_1 with T_3, $T_1 T_3$ with T_2, and finally $T_1 T_3 T_2$ with T_4.

We must now determine, in logarithmic time, which transposition should be applied to a permutation. Based on Lemma 3, the *Query* procedure solves this problem.

Lemma 3. *[1] Let b_i and b_j be two black edges in an unoriented cycle C, $i < j$. Let $\pi_k = \max_{i < m \leq j} \pi_m$, $\pi_\ell = \pi_k + 1$, then black edges b_k and $b_{\ell-1}$ belong to the same cycle, and pair $b_k, b_{\ell-1}$ intersects pair b_i, b_j.*

The procedure $Query(\pi, i, j)$, described in Algorithm 2 finds a pair of reality edges in $G(\pi)$ that intersects b_i, b_j in time $O(\log n)$. Every split in step 2 is performed in time $O(\log n)$.

Algorithm 2. $Query(\pi, i, j)$

---

    input   : permutation π, integers i and j
1 Let T be the permutation tree of π
2 Split T, into three permutation trees, T_1, T_2 and T_3, corresponding to
   $[\pi_0, \pi_1, \cdots, \pi_i]$, $[\pi_{i+1}, \cdots, \pi_j]$, and $[\pi_{j+1}, \cdots, \pi_n, \pi_{n+1}]$, respectively.
3 Let $\pi_k = root(T_2)$. (the largest element in the interval $\pi_{i+1}, \cdots, \pi_j$)
4 Let $\pi_\ell = \pi_k + 1$
5 Return the pair k, $\ell - 1$ (by Lemma 3, $b_k, b_{\ell-1}$ intersects b_i, b_j)

The *Query* procedure is the method used in Hartman and Shamir's [10] 1.5-approximation algorithm to find a $(3, 2)$-sequence that affects a pair of intersecting or interleaving cycles.

Firoz *et al.* [8] suggested the use of the permutation tree data structure to reduce the running time of Algorithm 1 to $O(n \log n)$, but in Sect. 3 we show that this strategy, in the manner proposed by Firoz *et al.*, may fail to extend a full configuration to 9 cycles.

3 The Use of the Permutation Tree by Firoz et al.

Firoz *et al.* [8] state that step 9 in Algorithm 1 can be done in $O(\log n)$ time. To do so, they categorized sufficient extensions of a configuration A into *type 1 extensions* – those that add a cycle that closes an open gate – and *type 2 extensions* – those that extend a full configuration by adding a cycle C such that $A \cup \{C\}$ has at most one open gate.

A type 1 extension can be performed in logarithmic time with a $Query(\pi, i, j)$, where b_i, b_j form an open gate. For a type 2 extension, since there are no open gates, Firoz *et al.* claim that it is sufficient to perform queries with every pair of reality edges that belong to the same cycle in the configuration that is being extended. Example 1 shows that this strategy is flawed.

Example 1. *Consider the permutation $\pi = [0\,10\,9\,8\,7\,1\,6\,11\,5\,4\,3\,2\,12]$, whose breakpoint graph is depicted in Fig. 1. It is a simple permutation having only* *unoriented 3-cycles. We mark all the cycles* $C_1 = \langle b_0, b_2, b_4 \rangle$, $C_2 = \langle b_1, b_3, b_6 \rangle$, $C_3 = \langle b_5, b_8, b_{10} \rangle$ *and* $C_4 = \langle b_7, b_9, b_{11} \rangle$, *let* $A = \{C_1\}$ *be the configuration and try to sufficiently extend* A *(step 9 in Algorithm 1) using the method proposed by Firoz et al.:*

1. *Configuration* A *has three open gates* (b_0, b_2), (b_2, b_4), (b_4, b_0). *We execute* $Query(\pi, 0, 2)$ *(or, alternatively* $Query(\pi, 2, 4)$, *with the same end result), wich returns the pair* b_1, b_6, *in cycle* $\langle b_1, b_3, b_6 \rangle$. *Therefore, we add this cycle to the configuration* A, *which becomes* $A = \{C_1, C_2\}$.
2. *Configuration* A *has no more open gates. We must execute* $Query(\pi, i, j)$ *for every pair of elements* b_i, b_j *in the same cycle of the configuration such that* $i < j$; *it is easy to observe that Query will returns a pair that is already in* A. *So far, Firoz et al.'s method has failed to extend* A.
3. *Configuration* A *is not a component, therefore we unmark all the cycles in* A.
4. *The marked cycles are now* C_3 *and* C_4. *If we consider* $A = \{C_3\}$ *or* $A = \{C_4\}$, *using Firoz et al.'s method only allows us to extend* A *as far as* $\{C_3, C_4\}$. *Again,* A *is not a component.*

Therefore, Firoz et al.'s method fails to find the component $\{C_1, C_2, C_3, C_4\}$.

The permutation in Example 1, although having only one small component, clarifies the problem with Firoz *et al.*'s strategy for type 2 extensions. The same problem happens for sufficient configurations with more than nine cycles, such as:

$$\sigma = [\mathbf{0}\, 25\, 24\, 23\, 22\, 1\, 21\, 26\, 20\, 19\, 18\, 2\, 17\, 27\, 16\, 15\, 14\, 3\, 13\, 28\, 12\, 11\, 10\, 4\, 9\, 29\, 8\, 7\, 6\, 5\, \mathbf{30}].$$

Fig. 3 displays the breakpoint graph of σ, which consists of five pairs of interleaving 3-cycles, defining a single component that is not small. In this case, starting the *Query* procedure with a configuration that has a unique cycle, it is possible only to perform a single type 1 extension, but any subsequent type 2 extension fails.

By Lemma 1, every configuration of nine cycles has an 11/8-sequence. Fig. 3 illustrates a case where there exists an 11/8-sequence. Starting the *Query* from a configuration with a unique cycle, the procedure returns five pairs of interleaving 3-cycles, each pair would correspond to a bad small component, as illustrated in Fig. 4. Notice that according to Step 18 of Algorithm 1, if a permutation contains only bad small components, then an 11/8-sequence is returned, see Fig. 5 for the breakpoint graph of γ:

$$\gamma = [\mathbf{0}\, 5\, 4\, 3\, 2\, 1\, 6\, 11\, 10\, 9\, 8\, 7\, 12\, 17\, 16\, 15\, 14\, 13\, 18\, 23\, 22\, 21\, 20\, 19\, 24\, 29\, 28\, 27\, 26\, 25\, \mathbf{30}],$$

a permutation which Algorithm 1 may wrongly consider instead of σ. However, Algorithm 1 does not have a rule to deal with five pairs of interleaving cycles where, for each pair, there exists another intersecting cycle, as in Fig. 3.

Notice that the permutations of Example 1 and the above permutation σ are examples belonging to a family of permutations such that the type 2 extension fails. Actually, an infinite family can be constructed as follows: Let k be any integer grater than or equal to 2, and let $f(i)$ be the sequence of six integers

$$i \; 5k-4i \; 5k+i \; 5k-4i-1 \; 5k-4i-2 \; 5k-4i-3.$$

Consider σ^k a permutation of $6k - 1$ elements defined as:

$$[\mathbf{0} \; 5k \; 5k-1 \; 5k-2 \; 5k-3 \; f(1) \; f(2) \ldots f(k-1) \; k \; \mathbf{6k}],$$

whose breakpoint graph has a similar structure to that in Figs. 1 and 3. If we start from a configuration having any one of the cycles, it is impossible to extend it past a configuration of more than two cycles using Firoz et al.'s approach.

Fig. 3. Breakpoint graph of a permutation σ for which Firoz et al. method fails. Note that σ has 10 cycles, and that σ is obtained from the permutation in Fig. 1.

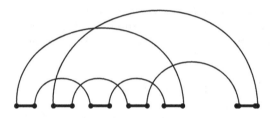

Fig. 4. A configuration that is not maximal returned by the *Query* on σ

Fig. 5. The breakpoint graph of permutation γ

4   Conclusions

The goal of this paper is to show an infinite family of permutations that invalidate the correctness of Elias and Hartman's algorithm with the Firoz et al.'s strategy. We investigated Elias and Hartman's 1.375-approximation algorithm for the SBT problem and the claimed strategy to improve the running time of such algorithm from $O(n^2)$ down to $O(n \log n)$. Clearly, the use of Feng and Zhu's permutation tree is not enough. We are currently investigating alternative counter-examples that may suggest how hard it could be to correct Firoz et al.'s approach.

References

1. Bafna, V., Pevzner, P.A.: Sorting by transpositions. SIAM J. Disc. Math. 11, 224–240 (1998)
2. Bulteau, L., Fertin, G., Rusu, I.: Sorting by transpositions is difficult. SIAM J. Discrete Math. 26(3), 1148–1180 (2012)
3. Cunha, L.F.I.: Limites para Distância e Diâmetro em Rearranjo de Genomas por Transposições. Master dissertation, Programa de Engenharia de Sistemas e Computação – COPPE/UFRJ, Brazil (2013)
4. Cunha, L.F.I., Kowada, L.A.B., de A. Hausen, R., de Figueiredo, C.M.H.: Transposition diameter and lonely permutations. In: de Souto, M.C.P., Kann, M.G. (eds.) BSB 2012. LNCS (LNBI), vol. 7409, pp. 1–12. Springer, Heidelberg (2012)
5. Dias, U., Dias, Z.: An improved 1.375-approximation algorithm for the transposition distance problem. In: Proceeding of the 1st ACM International Conference on Bioinformatics and Computational Biology (ACM-BCB 2010), pp. 334–337. ACM, Niagara Falls (2010)
6. Elias, I., Hartman, T.: A 1.375-approximation algorithm for sorting by transpositions. IEEE/ACM Trans. Comput. Biol. Bioinformatics 3(4), 369–379 (2006)
7. Feng, J., Zhu, D.: Faster algorithms for sorting by transpositions and sorting by block interchanges. ACM Trans. Algorithms 3(3) (2007) 1549–6325
8. Firoz, J.S., Hasan, M., Khan, A.Z., Rahman, M.S.: The 1.375 approximation algorithm for sorting by transpositions can run in $O(n \log n)$ time. J. Comput. Biol. 18(8), 1007–1011 (2011)
9. Hannehalli, S., Pevzner, P.: Transforming cabbage into turnip: polynomial algorithm for sorting signed permutations by reversals. J. ACM 46, 1–27 (1999)
10. Hartman, T., Shamir, R.: A simpler and faster 1.5-approximation algorithm for sorting by transpositions. Inf. Comput. 204(2), 275–290 (2006)
11. Hausen, R.A., Faria, L., Figueiredo, C.M.H., Kowada, L.A.B.: Unitary toric classes, the reality and desire diagram, and sorting by transpositions. SIAM J. Disc. Math. 24(3), 792–807 (2010)
12. Kaplan, H., Verbin, E.: Efficient data structures and a new randomized approach for sorting signed permutations by reversals. In: Baeza-Yates, R., Chávez, E., Crochemore, M. (eds.) CPM 2003. LNCS, vol. 2676, pp. 170–185. Springer, Heidelberg (2003)
13. Kowada, L.A.B., de A. Hausen, R., de Figueiredo, C.M.H.: Bounds on the transposition distance for lonely permutations. In: Ferreira, C.E., Miyano, S., Stadler, P.F. (eds.) BSB 2010. LNCS (LNBI), vol. 6268, pp. 35–46. Springer, Heidelberg (2010)
14. Lopes, M.P., Braga, M.D.V., de Figueiredo, C.M.H., de A. Hausen, R., Kowada, L.A.B.: Analysis and implementation of sorting by transpositions using permutation trees. In: Norberto de Souza, O., Telles, G.P., Palakal, M. (eds.) BSB 2011. LNCS (LNBI), vol. 6832, pp. 42–49. Springer, Heidelberg (2011)
15. Sankoff, D., Leduc, G., Antoine, N., Paquin, B., Lang, B.F., Cedergren, R.: Gene sort comparisons for phylogenetic inference: evolution of the mitochondrial genome. Proc. Natl. Acad. Sci. 89(14), 6575–6579 (1992)

ncRNA-Agents: A Multiagent System for Non-coding RNA Annotation

Wosley Arruda[1], Célia G. Ralha[1], Tainá Raiol[2], Marcelo M. Brígido[2],
Maria Emília M.T. Walter[1], and Peter F. Stadler[3]

[1] Department of Computer Science, University of Brasília, Brazil
[2] Laboratory of Molecular Biology, Institute of Biology, University of Brasília, Brazil
[3] Department of Computer Science and Interdisciplinary Center for Bioinformatics, University of Leipzig, Germany

Abstract. In recent years, non-coding RNAs (ncRNAs) have been focus of intensive research. Since the characteristics and signals of ncRNAs are not entirely known, researchers use different computational tools together with their biological knowledge to predict potential ncRNAs. In this context, this work presents a multiagent system to annotate ncRNAs based on the output of different tools, using inference rules to simulate biologists' reasoning. Experiments with real data of fungi allowed to identify novel putative ncRNAs, which shows the usefulness of our approach.

1 Introduction

Since 1990, important roles of the RNA molecules have been identified, besides the well defined functions of messenger RNA (mRNA), transporter RNA (tRNA) and ribosomal RNA (rRNA), all involved in protein synthesis. Non-coding RNAs (ncRNAs), those RNAs not coding for proteins [6,25], present specific spatial conformation that allow them to play regulatory roles in an extensive variety of biological reactions and processes, e.g., translation initiation, level control of mRNA, stem-cell maintenance, brain developing, metabolism regulation, support to protein transport, nucleotide edition, imprinting regulation and chromatin dynamics [22].

On one side, although intensive efforts worldwide, biological and computational methods are not yet capable to easily identify and classify ncRNAs, directly affecting the annotation of these transcripts. From a biological point of view, ncRNAs are characterized by its transcription and absence of translation into proteins, and RNAs presenting very different nucleotide sequences (primary sequences) but similar spatial conformations (secondary structure) perform the same cellular functions. Therefore, ncRNAs should be characterized by their secondary structures and not only by their primary sequences. From a computational point of view, ncRNAs can not be identified and classified by homology tools, extensively and efficiently used to annotate protein coding genes (e.g., BLAST), and methods designed to predict secondary from primary structure are commonly used to annotate ncRNAs [13,6]. In this context, biologists use

J.C. Setubal and N.F. Almeida (Eds.): BSB 2013, LNBI 8213, pp. 136–147, 2013.

different tools to annotate sequences that appear to be ncRNAs, together with their knowledge, which is a reasoning intensive process.

On the other side, a multiagent system (MAS) is characterized by the distribution of intelligence among autonomous entities called agents, which interact to reach individual or collective goals. In order to do that, the agents in a MAS have to negotiate in a cooperative way, coordinating and joining efforts to reach the objectives, normally not accomplished by a single agent [31].

In this context, this work has the objective of presenting ncRNA-Agents, a MAS to support ncRNA annotation, using inference rules to simulate the biologists' reasoning to combine the output of different tools and data bases [27]. As far as we know, there are no computational tools simulating the biologists' reasoning to annotate ncRNAs.

This work is divided into six sections. In Section 2, challenges to annotate ncR-NAs are firstly discussed, together with commonly used tools and data bases. After, a general classification for ncRNA annotation tools is proposed. In Section 3, concepts and tools to implement MAS are introduced. In Section 4, the ncRNA-Agents architecture and some implementation details are detailed. In order to validate ncRNA-Agents, two experiments with real data of fungi are analized in Section 5. Finally, this work is concluded in Section 6.

2   Challenges to Annotate ncRNAs

Methods to identify and classify ncRNAs use different strategies, based on ncRNA characteristics supported by *in silico* and experimental findings. Some criteria are commonly used: a number of known ncRNA classes do not present long ORFs; their sequences have unexpected stop codons; RNAs usually present conservation in their secondary (spatial) structure and rarely in their primary (DNA sequences) structure; some known ncRNAs have complex tridimensional structures, and catalyst or structural functions. These characteristics prevent to use traditional similarity based tools to predict proteins [25]. Methods using biological hints are also extensively used to predict ncRNAs: codons, synonymous and non-synonymous substitutions, and minimum folding energy [21].

One important problem when predicting ncRNAs is the lack of available experimental information, confirming *in silico* prediction. Although computational predictions are useful, confirmation needs biological experiments, such as RNAseq of small RNAs, real-time PCR and gene deletion or knock out experiments. But the absence of translation is not conclusive, since predicted ncRNA may not be transcribed or translated when some environmental or physiological conditions happen. Therefore, to predict ncRNAs, biologists use different bioinformatics tools, with distinct methods, and after make a combined analysis of all these information to decide if a sequence is a potential ncRNA. We have three main problems when annotating ncRNAs: secondary structure prediction [14], secondary structure comparison [13,6] and ncRNA identification and classification [10].

2.1   Tools and Data Base to Annotate ncRNAs

In this section, we describe first some computational tools and after data bases [30] commonly used to identify and classify ncRNAs.

BLAST (*Basic Local Alignment Search Tool*) [1] is extensively used to predict proteins. Although Blast can successfully detect some specific ncRNA families (e.g., snoRNAs), in general, only nucleotide comparisons is not enough, since many ncRNA classes are very different regarding to the primary sequences but exhibit a common secondary structure. **Infernal** (INFERence of RNA Alignment) [5] identifies ncRNAs looking for secondary structures. It builds profiles of RNA consensus spatial structure, known as covariance models (CMs), in order to find similarities between the investigated sequence and the consensus secondary structure of each one of the RNA families stored in the Rfam data base [11]. **tRNAscan** [19] is considered one of the more precise tRNA predictors. It combines three programs: two tRNA predictors and a covariance model [5] previously trained with tRNA sequences. The execution of these three programs results is a tRNA identifier presenting high sensibility (99 − 100%) and specificity (with a false positive rate less than 0.00007 by Mb) in a reasonable velocity. **Portrait** [2] identifies ncRNAs of not complete transcriptomes of yet not entirely characterized species, based on Support Vector Machine (SVM). The result of Portrait is a probability indicating the likelihood of a transcript to be non protein coding. **Vienna** [13] is a set of packages used to generate or to compare RNA secondary structures. Folding in this tool uses prediction algorithms based on RNA free energy and the probability of base pairing [21]. Particularly, RNAfold [13,14] package is based on the hypothesis that an RNA molecule is folded in a more stable thermodynamics structure, the one presenting the minimum free energy. **RNAmmer** [16] predicts rRNAs using the 5S ribosomal and the European ribosomal RNA data base to generate many structural alignments, which are used to build Markov chain libraries.

Now, we briefly describe commonly used ncRNA data bases, created from experimental and computational data [30]. **NONCODE** [18] includes many different types of ncRNAs, except tRNAs and rRNAs. More than 80% of the input in NONCODE are experimentally confirmed. We used NONCODE version v3.0, with more than 411, 552 ncRNAs. **RNAdb** [24] contains sequences and annotation of thousands of mammalian ncRNAs, but a large number of them do not have known biological functions. **miRBase** [12] is a data base of microRNAs. **snoRNA** database [17] contains human snoRNAs of both types, H/ACA and C/D box. **fRNAdb** [23] integrates a set of data bases, including NONCODE and RNAdb. **Rfam** [11] is a curated data base containing information about ncRNA families, presenting two classes of data: CM profile and seed alignments for each ncRNA family. In this work, we used Rfam 11.0, with 2, 208 families.

2.2   A Proposal to Classify ncRNA Annotation Tools

We propose a general classification for computational tools that annotate ncRNAs, in three groups, described as follows: (i) **Homology**: in this group, ncRNAs are predicted using alignment among sequences. These comparisons depend on the quality of the annotation stored in data bases, since this quality affects ncRNA classification. BLAST [1] and Infernal [6] are two extensively used tools, both using the concept of similarity (the greater the similarity the greater the chance that both sequences have conserved the same function). In general, BLAST does not produce good results to annotate ncRNAs, being Infernal more sensitive and specific to identify hundreds of different types of ncRNAs; (ii) **Class prediction**: in this group, ncRNAs are identified using methods based on machine learning. In the supervised learning paradigm, we take a set of known ncRNAs and a set of known proteins, computing *ab initio* characteristics for these sequences, in order to create a model to predict ncRNAs. Examples are PORTRAIT [2] and DARIO [7]; and (iii) *De novo* **model**: in this group, ncRNAs are predicted using models distinct from *homology* and *class prediction*, e.g., the Vienna [13] thermodynamic model.

3   MAS and Inference Rules

In this section, we briefly describe MAS, as well as inference rules, respectively used for knowledge representation and reasoning. First, we present the concept of an agent. According to Russell and Norvig [29], an agent is an entity that interacts with the environment, perceiving it through its sensors and acting using actuators. An agent has two important characteristics: it is capable to act autonomously, taking decisions leading to the satisfaction of its objectives; and it is capable to interact with other agents through human inspired social interaction protocols. We may cite as agents functionalities: coordination, cooperation, competition and negotiation.

A MAS includes many homogeneous or heterogeneous agents interacting and working together. Each agent operates asynchronously related to other agents [31]. In order to have agent communication and interaction we need to have an adequate MAS infrastructure with specific agent language and interaction protocol.

JADE (Java Agent DEvelopment Framework) [3] is a commonly used framework to develop MAS, since it follows the patterns of FIPA (Foundation for Intelligent Physical Agents), an international organization responsible to define patterns for the development of agent technologies. JADE allows and facilitates the development of agents using Java language since it has many already defined functions, uses ACL (Agent Communication Language), and has ready to use interaction protocols (e.g., contract net). JADE presents a visual interface to monitor agents' execution that helps to control these agents' life cycle, having also many built-in resources, e.g., the directory facilitator (yellow pages) and the agent controller manager (white pages). In this work, we have used JADE to implement the ncRNA-Agents prototype.

In intelligent agents, reasoning skills can be implemented using inference rules, which also allows to represent knowledge that will be manipulated by the inference engine. Drools [4] is an open source rule engine that allows to build a knowledge base and make inferences based in patterns. Drools interacts with Java and the knowledge is obtained from a set of declarative rules. A Drools' rule has one or more condition (or facts) leading to one or more actions (or consequences). Basically, the Drools inference engine offers the possibility of using a forward chaining or a backward chaining method as investigation approaches. The inference algorithm of Drools is RETE [9], which is adapted to object oriented systems. In order to integrate the inference rule engine to JADE framework we have used Drools.

4   ncRNA-Agents Architecture and Prototype

In this section, we first present the ncRNA-Agents architecture, and after details of the implemented prototype. This ncRNA annotation system using MAS was inspired in previous work of our group [26,27].

4.1   Architecture Proposal

Figure 1 presents the ncRNA-Agents architecture with four different layers: (i) **interface layer**: receives from a user one sequence (or a sequence file) in FASTA format, together with the required annotation tools. The system will return the RNA annotation (a potential ncRNA or not) of each sequence, along with the used tools and corresponding reasoning; (ii) **conflict resolution layer**: receives the user request and decides the better recommendations based on suggestions received from the collaborative layer, sending them to the interface layer; (iii) **collaborative layer**: execute tools (the user can choose the tools among the ones presented in the ncRNA-Agents initial page) to annotate ncRNAs, sending the obtained results to the conflict resolution layer; and (iv) **physical layer**: stores different data bases, as those presented in Section 2.1.

In Figure 1, note that the collaborative layer present different levels of agents: manager and analyst agents. **Manager agents** are responsible for filtering the suggestions sent by the analyst agents. We defined four types of manager agents, three of them following the classification of the computational methods to annotate ncRNAs proposed in Section 2.2: (i) **homology manager agent**: coordinates agents working on tools based on homology, e.g., BLAST [1] and Infernal [6]; (ii) **class prediction manager**: coordinates agents working on tools based on machine learning, e.g., Portrait [2]; (iii) *de novo* **manager**: coordinates agents working on tools that do not use reference organisms, e.g., RNAfold [14] from Vienna package; and (iv) **seeker manager**: coordinates agents created to refine analyses of the other manager agents (mainly removing false positives), simulating the reasoning of a biologist when analyzing the results of the required programs and data bases. It also gives flexibility to the architecture, since it can be adapted according to the purposes of each project. Finally, **analyst agents** are responsible for executing specific tools to annotate ncRNAs. Each agent (created from a manager agent request) parses the output file generated by the tool controlled by this agent. This analysis is returned as an annotation to its corresponding manager agent. Particularly, analyst agents

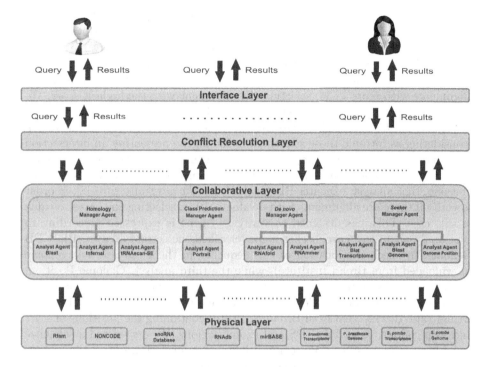

Fig. 1. ncRNA-Agents architecture

can be created in seeker manager to improve annotation, e.g., to check if a ncRNA candidate belongs to intronic or intergenic region, or to build a multiple alignment for investigating conserved domains among related organisms. This manager allows to remove false positives, e.g, seeker manager can verify if a predicted snoRNA does not belong to an exonic region.

4.2   Prototype Implementation

A prototype was created with four manager agents: homology, class prediction, *de novo* and seeker. The ncRNA-Agents system was written to be time efficient using threads executed in parallel. It was implemented with JADE version 4.1 [3]. *Eclipse SDK* version Helios 2012 was used as the development environment together with the web server apache tomcat 7.0. *JADE* was adopted for a couple of reasons. It is a free software distributed under *LGPL* license, and *Java* language allows portability. Since it is a ready to use plataform, it is not necessary to implement agent funcionalities, agent management ontologies and message transport mechanism.

ncRNA-Agents uses *Drools* version 5.5.0 for the agents reasoning [4]. We defined the biological knowledge using declarative rules according to the parameters defined for the experiments. Rules for homology manager were created for choosing, among the results of the three analysts (Blast, Infernal and tRNAscan), the better annotation, which is sent to the conflict resolution layer. tRNAscan analyst verifies if there are any good alignments, selecting the alignment with the highest score. Blast analyst verifies if there are alignments with e-value $\leq 10^{-5}$ (adopted by the biologists, but this value is an easily changeable system parameter), selecting the lower e-value in this case. Infernal analyst verifies if there are alignments with score ≥ 34, selecting the alignment with the highest score. We note that the analyst agents execute in parallel.

We called *potential ncRNA* an expressed sequence located in an intronic region (considering as intronic those sequences inside a gene presenting an intersection of at most 50% with an exon) or in an intergenic region. To identify such a ncRNA, the rules for seeker manager were designed using three analysts. The first one, Blat analyst, executes Blat with *P. brasiliensis* transcripts [8], in order to verify if there are alignments with e-value $\leq 10^{-5}$, selecting the lowest e-value in this case. This was designed to verify if a gene is expressed (according to Blat analyst). In this case, seeker manager asked the second analyst, Blast PB Genome, to check the location of this sequence in the *P. brasiliensis* genome (downloaded from Broad Institute), while the third analyst, Genome Position, was in charge to verify if the sequence occurs in an intronic or an intergenic region, as described above. Genome Position analyst takes as input a file in .gtf format (also downloaded from Broad Institute) containing the positions of start and stop codons, as well as CDS and exonic regions. We considered a sequence to be located in an intergenic region if it was not located in one of the regions reported in this file but was close enough to a gene (≤ 1.000 bp).

Finally, the conflict resolution layer (CR layer) rules were designed as follows. The first decision is about annotating the sequence as tRNA, if the homology manager sends an annotation identified by tRNAscan analyst. If it is not the case, CR layer decides if the sequence can be annotated as rRNA, if *de novo* Manager sends an annotation identified by RNAmmer analyst. Both tools are considered reliable to annotate tRNAs and rRNAs, respectively. If the sequence could not be annotated as tRNA or rRNA, CR layer verifies if homology manager sent a recommendation based on Infernal analyst or Blast analyst (with databases snoRNA, RNAdb, NONCODE and mirBASE, in this order), sending it to the interface layer if it is the case.

If a recommendation could not be found by homology manager, CR layer verifies if class prediction manager sends a recommendation. If the probability of being a ncRNA is $< 70\%$, the recommendation of the Portrait analyst is "not annotated by Portrait". On the other hand (probability $\geq 70\%$), CR layer validates this with other analyses, as follows. The first verification is done by *de novo* manager. It executes RNAfold analyst, which indicates as potential ncRNA those sequences with *Minimum Free Energy (MFE)* ≤ -5.00 *kcal/mol*. If it was not the case, the recommendation is "not annotated by RNAfold". Otherwise, CR layer calls seeker manager to check the three aspects described above: (i) verifies if the RNA is expressed in the *P. brasiliensis* trancriptome (Blat Analyst); (ii) finds where this RNA is located in the chromosomes of the *P. brasiliensis* genome (Blast Analyst); and (iii) verifies if this sequence is located in intron, exon or intergenic region.

5   Results

To validate the prototype, we developed two experiments with real data of two fungi: *Paracoccidioides brasiliensis* [8] and *Schizosacharomyces pombe* [20].

Detecting ncRNAs in *Paracoccidioides brasiliensis*

First, 6, 022 sequences from *P. brasiliensis* transcriptome [8] were submitted to Blast with SwissProt data base, to annotate proteins. Sequences not identified as proteins were submitted as input to ncRNA-Agents, noting that sequences annotated as "hypothetical protein", "predicted protein", "potential protein" and "unknown sequence" were also included in this input file. For the experiment, BLAST was executed with data bases snoRNA, RNAdb, NONCODE and mirBASE. Infernal was executed with Rfam 11.0 data base, and the *de novo* tools tRNAscan and Portrait were both executed. We also used Blat with *P. brasiliensis* transcripts, Blast with *P. brasiliensis* genome, and a script to verify if a sequence was located in intergenic or intronic regions.

Table 1. ncRNAs in *P. brasiliensis* identified by ncRNA-Agents. Column "Genome Position" indicates if a sequence is in intronic or intergenic region (in this case, the closest gene is indicated).

Sequence Name	SuperContig	Initial Position	Final Position	ncRNA Length (bp)	Strand	Genome Position
Pb_ncRNA_1	Supercontig_1.13	895440	895118	322	-	Intergenic: After PAAG_05248
Pb_ncRNA_2	Supercontig_1.48	5192	5578	386	+	Intron
Pb_ncRNA_3	Supercontig_1.3	248422	248850	428	+	Intergenic: Before PAAG_01391
Pb_ncRNA_4	Supercontig_1.109	3734	3518	216	-	Intron
Pb_ncRNA_5	Supercontig_1.58	25692	26297	605	+	Intergenic: After PAAG_08977
Pb_ncRNA_6	Supercontig_1.39	84839	84453	386	-	Intron
Pb_ncRNA_7	Supercontig_1.33	1559	1734	175	+	Intergenic: Before PAAG_08374
Pb_ncRNA_8	Supercontig_2.66	478	100	378	-	Intergenic: Before PAAG_12001
Pb_ncRNA_9	Supercontig_1.9	14977	15094	117	+	Intergenic: Before PAAG_03665

We found 9 potential ncRNAs, as shown in Table 1. Pb_ncRNA_1, located into a intergenic region, was identified as the Fungi_SRP (Fungal signal recognition particle RNA), the RNA component of the signal recognition particle (SRP) ribonucleoprotein complex [28]. SRP RNA, also known as 7SL, 6S, ffs, or 4.5S RNA, participates in the coordination of protein traffic to cellular membranes. The RNA and protein components of the ribonucleoprotein complex are highly conserved in all domains of life, but varying the molecule size and the number of RNA secondary structure elements. The eukaryotic SRP consists of a 300-nucleotide 7S RNA and six proteins: SRPs 72, 68, 54, 19, 14 and 9. Although we could not classify Pb_ncRNA_3, it presented a high similarity with the small ncRNA AFU_254 (sRNA Afu-254) found in *Aspergillus fumigatus*, a pathogenic filamentous fungus responsible for infections worldwide, suggesting a conservation of this potential small ncRNA between these two pathogenic fungi.

Detecting ncRNAs in *Schizosacharomyces pombe*

Small RNA-Seq data of *S. pombe* was extracted from EMBL-EBI (experiment E-MTAB-1154). After filtering to remove low quality reads, mapping was performed using Segehmel [15], taking as reference the *S. pombe* genome (downloaded from BROAD Institute). Next, a Perl script identified genome continuous regions presenting at least five mapped reads, and another Perl script found the location of each region in the *.bed* file generated by mapping. The annotation followed the same reasoning designed for *P. brasiliensis*, except for the seeker

Table 2. Some ncRNAs identified in the three chromosomes of *S. pombe* by ncRNA-Agents. Column "Sequence Name" shows the chromosome (I, II, III) and column "Genome Position" presents the sequence location (intron, exon, ncRNA exon, and the related gene as annotated in the .gtf file)

Sequence Name	Region Name	Initial Position	Final Position	ncRNA Length (bp)	Genome Position	Strand
Sp_ncRNA_I1	region90	98789	98839	50	Intron Gene: SPAC1F8.05	+
Sp_ncRNA_I2	region553	365148	365236	88	tRNA exon Gene: SPATRNAPRO.01	+
Sp_ncRNA_I3	region707	431217	431301	84	ncRNA Exon Gene: SPNCRNA.645	+
Sp_ncRNA_I4	region1059	617512	617563	51	Intron: After Gene: SPAC1F3.02c	+
Sp_ncRNA_I5	region2501	1421144	1421233	89	Exon Gene: SPAC20G8.10c	+
Sp_ncRNA_II1	region488	355913	356005	92	ncRNA Exon Gene: SPNCRNA.1343	+
Sp_ncRNA_II2	region925	570150	570239	89	rRNA Exon Gene: SPBRRNA.30	+
Sp_ncRNA_II3	region2686	1599696	1599799	103	tRNA Exon Gene: SPBTRNAGLY.07	+
Sp_ncRNA_II4	region2914	1774859	1774940	81	Exon Gene: SPBC18H10.03	+
Sp_ncRNA_II5	region6626	3944593	3944692	99	snoRNA Exon Gene: SPBC26H8.02c	+
Sp_ncRNA_III1	region840	489561	489616	55	Intron: After Gene: SPCC970.11c	+
Sp_ncRNA_III2	region1001	583418	583614	196	ncRNA Exon Gene: SPNCRNA.467	+
Sp_ncRNA_III3	region1891	1071041	1071190	149	tRNA Exon Gene: SPCTRNALEU.11	+
Sp_ncRNA_III4	region1910	1102800	1102922	122	tRNA Exon Gene: SPCTRNAGLU.10	+
Sp_ncRNA_III5	region3149	1863119	1863215	96	Exon: After Gene: SPCC1223.10c	+

Fig. 2. RNAfold analysis of (a) Sp_ncRNA_I4 (C/D Box), (b) Sp_ncRNA_I5 (tRNA), (c) Sp_ncRNA_II5 (H/ACA Box), (d) Sp_ncRNA_III5 (C/D Box), and (e) region 708 (annotated as "snRNA exon gene SPSNRNA.04" by seeker manager)

manager, which only verified if the predicted ncRNA was located in exon, intron or intergenic region (Blat and Blast analysts were not used here). The first script was executed in the three chromosomes of *S. pombe*, having identified respectively 9, 566, 7, 565 and 4, 126 regions. These regions were submitted to ncRNA-Agents, which identified 52 putative ncRNAs, all in strand "+". General features of some of those ncRNAs are shown in Table 2.

Sequences not annotated as ncRNAs by homology, class prediction and *de novo* managers were submitted only to seeker manager. This allows to find other 2,493 putative ncRNAs (728 in exons and 1,765 in introns) using the annotation of the reference .gtf file: (i) in chromosome I, 327 in exons (7 tRNAs, 2 snoRNAs, 1 snRNA, 317 putative ncRNAs) and 834 in introns; (ii) in chromosome II, 276 in exons (17 tRNAs, 9 rRNAs, 4 snRNAs, 246 putative ncRNAs) and 618 in introns; and (iii) in chromosome III, 125 in exons (9 tRNAs, 4 rRNAs, 1 snoRNA, 111 putative ncRNAs) and 313 in introns.

It is important to note that, in both cases (predicted ncRNAs obtained from all the managers working together and those by seeker manager only), to be predicted as putative ncRNAs, a sequence had to present folding free energy $<$ -5.00 kcal/mol (according to RNAfold), indicating likely stable conformational structures. Figure 2 shows some examples.

6   Conclusion

In this article, we presented ncRNA-Agents, a multiagent system to support ncRNA annotation. The proposed architecture allows the execution of intelligent agents cooperating in a heterogeneous and dynamic environment. Different tools and data bases as well as inference rules simulating biologists' reasoning can be specified according to the purposes of each project. In ncRNA-Agents, agents are specialized in different tasks, and can act independently using distinct inference rules. A prototype was implemented using JADE framework and Drools as inference engine for reasoning rules in a web environment. To validate ncRNA-Agents, we executed two experiments to detect ncRNAs in two fungi, *Paracoccidioides brasilienses* and *Schizosacharomyces pombe*. We obtained novel potential ncRNAs for both fungi, showing the usefulness of our approach.

We believe that ncRNA-Agents can strongly help to improve the quality of ncRNA annotation process, considering that massively parallel automatic sequencers produce billions of small fragments. As far as we know, there is no system simulating the biologists reasoning with the integration of different tools and data bases through a multiagent approach as we presented.

Next steps include to improve the expertise of the agents in ncRNA-Agents, using more specific reasoning mechanisms. The inference rules could also be extended, producing a more intensive knowledge based tool. In this direction, we would like to identify, if the annotations are not conclusive enough, a set of new tools to support annotation, e.g., if a potential RNA belongs to a known ncRNA family using a multiple alignment with known and conserved RNAs in related organisms, and figures showing a putative ncRNA folding. Data mining methods could also improve precision in the annotation process. In the collaborative layer, other analyst agents (corresponding to other tools) can be easily included in each manager agent. Particularly, in the class prediction manager, a tool to classify ncRNAs could be included, besides Portrait that only indicate if a sequence is a potential ncRNA. Besides, implementation aspects should be improved allowing a reproducible and portable annotation tool that may be integrated to different annotation systems. Distributed implementation of the agents (besides threads) could reduce time execution, improving the use of ncRNA-Agents in large-scale sequencing projects.

References

1. Altschul, S.F., et al.: Basic local alignment search tool. Journal of Molecular Biology 215(3), 403–410 (1990)
2. Arrial, R.T., Togawa, R.C., Brigido, M.M.: Screening non-coding RNAs in transcriptomes from neglected species using PORTRAIT: case study of the pathogenic fungus *Paracoccidioides brasiliensis*. BMC Bioinformatics 10(1), 239 (2009)
3. Bellifemine, F., Caire, G., Poggi, A., Rimassa, G.: JADE - a white paper. White Paper 3, TILAB - Telecom Italia Lab. 3, 6–19 (2003), http://jade.tilab.com/
4. Browne, P.: JBoss Drools Business Rules. Packt Publishing (2009)
5. Eddy, S.R., Durbin, R.: RNA sequence analysis using covariance models. Nucleic Acids Res. 22(11), 2079–2088 (1994)
6. Eddy, S.R., et al.: Non-coding RNA genes and the modern RNA world. Nature Reviews Genetics 2(12), 919–929 (2001), Infernal's user guide at,
http://infernal.janelia.org
7. Fasold, M., et al.: DARIO: a ncRNA detection and analysis tool for next-generation sequencing experiments. Nucleic Acids Res. 39, 1304–1351 (2011)
8. Felipe, M.S.S., et al.: Transcriptional profiles of the human pathogenic fungus *Paracoccidioides brasiliensis* in mycelium and yeast cells. Journal of Biological Chemistry 280(26), 24706–24714 (2005)
9. Forgy, C.L.: RETE: A fast algorithm for the many pattern/many object pattern match problem. Artificial Intelligence 19(1), 17–37 (1982)
10. Griffiths-Jones, S.: Annotating noncoding RNA genes. Annu. Rev. Genomics Hum. Genet. 8, 279–298 (2007)
11. Griffiths-Jones, S., et al.: Rfam: an RNA family database. Nucleic Acids Res. 31(1), 439–441 (2003), ftp://ftp.sanger.ac.uk/pub/databases/Rfam

12. Griffiths-Jones, S., et al.: miRBase: microRNA sequences, targets and gene nomenclature. Nucleic Acids Res. 34, D140–D144 (2006), miRBase database:
    `http://microrna.sanger.ac.uk/`
13. Hofacker, I.L., et al.: Fast folding and comparison of RNA secondary structures. Monatshefte für Chemie/Chemical Monthly 125(2), 167–188 (1994)
14. Hofacker, I.L., Fekete, M., Stadler, P.F.: Secondary structure prediction for aligned RNA sequences. Journal of Molecular Biology 319(5), 1059–1066 (2002)
15. Hoffmann, S., et al.: Fast mapping of short sequences with mismatches, insertions and deletions using index structures. PLoS Comput. Biology 5(9), e1000502 (2009)
16. Lagesen, K., et al.: RNAmmer: consistent and rapid annotation of ribosomal RNA genes. Nucleic Acids Res. 35(9), 3100–3108 (2007)
17. Lestrade, L., Weber, M.J.: snoRNA-LBME-db, a comprehensive database of human H/ACA and C/D box snoRNAs. Nucleic Acids Res. 34(suppl. 1), D158–D162 (2006)
18. Liu, C., et al.: NONCODE: an integrated knowledge database of non-coding RNAs. Nucleic Acids Res. 115(suppl. 1), D112–D115 (2005), NONCODE
    `http://www.noncode.org`
19. Lowe, T.M., Eddy, S.R.: tRNAscan-SE: a program for improved detection of transfer RNA genes in genomic sequence. Nucleic Ac. Res. 25(5), D955–D964 (1997)
20. Marguerat, S., et al.: Quantitative analysis of fission yeast transcriptomes and proteomes in proliferating and quiescent cells. Cell 151(3), 671–683 (2012)
21. McCaskill, J.S.: The equilibrium partition function and base pair binding probabilities for RNA secondary structure. Biopolymers 29(6-7), 1105–1119 (1990)
22. Michalak, P.: RNA world–the dark matter of evolutionary genomics. Journal of Evolutionary Biology 19(6), 1768–1774 (2006)
23. Mituyama, T., et al.: The functional RNA database 3.0: databases to support mining and annotation of functional RNAs. Nucleic Acids Res. 37(suppl. 1), D89–D92 (2009)
24. Pang, K.C., et al.: RNAdb 2.0: an expanded database of mammalian non-coding RNAs. Nucleic Acids Res. 35(suppl. 1), D178–D182 (2007), RNAdb -
    `http://jsm-research.imb.uq.edu.au/rnadb`
25. Pang, K.C., Frith, M.C., Mattick, J.S.: Rapid evolution of noncoding RNAs: lack of conservation does not mean lack of function. Trends in Genetics 22(1), 1–5 (2006)
26. Ralha, C.G., Schneider, H.W., Walter, M.E.M.T., Bazzan, A.L.C.: Reinforcement learning method for BioAgents. In: 11th Brazilian Symposium on Neural Networks, Sao Paulo, Brazil, pp. 109–114. IEEE (2010)
27. Ralha, C.G., Schneider, H.W., Walter, M.E.M.T., Brigido, M.M.: A multi-agent tool to annotate biological sequences. In: Filipe, J., Fred, A.L.N. (eds.) 3rd ICAART 2011. Agents, Rome, Italy, vol. 2, pp. 226–231. SciTePress (2011)
28. Rosenblad, M.A., Larsen, N., Samuelsson, T., Zwieb, C.: Kinship in the SRP RNA family. RNA Biol. 6, 508–516 (2009)
29. Russell, S.J., Norvig, P.: Artificial intelligence: A Modern Approach, 3rd edn. Prentice Hall (2010)
30. Soldà, G., et al.: An Ariadne's thread to the identification and annotation of non-coding RNAs in eukaryotes. Briefings in Bioinformatics 10(5), 475–489 (2009)
31. Wooldridge, M.J.: An introduction to multiagent systems. Wiley (2009)

Inference of Genetic Regulatory Networks Using an Estimation of Distribution Algorithm

Thyago Salvá, Leonardo R. Emmendorfer, and Adriano V. Werhli

Centro de Ciências Computacionais - C3
Programa de Pós-Gradução em Computacão
Universidade Federal do Rio Grande - FURG
Rio Grande, Brazil
{thyago.salva,leonardoemmendorfer,adrianowerhli}@furg.br

Abstract. Inference of Genetic Regulatory Networks from sparse and noisy expression data is still a challenge nowadays. In this work we use an Estimation of Distribution Algorithm to infer Genetic Regulatory Networks. In order to evaluate the algorithm we apply it to three types of data: **(i)** data simulated from a multivariate Gaussian distribution, **(ii)** data simulated from a realistic simulator, GeneNetWeaver and **(iii)** data from flow cytometry experiments. The proposed inference method shows a performance comparable with traditional inference algorithms in terms of the network reconstruction accuracy.

Keywords: Genetic Regulatory Networks, Estimation of Distribution Algorithm, Bayesian Networks.

1   Introduction

The area of Systems Biology and related fields have recently witnessed many discoveries and advances. These progresses are in great part due to the continuous increase in availability and diversity of molecular biology data. Usually, these studies are performed either considering single biological entities or the union of several such entities. It is now becoming clearer that the complexity of an organism is more related with the joint acting of the components rather than with the individual behaviour of its components. Therefore, the arrangement of biological components in networks is likely to play a pivotal role in crucial biological processes e.g. determining the development and sustainability of an organism. All these characteristics have prompted the necessity of studying these components in a holistic manner. Thus, the study of biological systems as biological networks is highly relevant and has the potential to help in deciphering the intricacy of living organisms.

As these complex networks are mainly unknown, one interesting approach is to reverse engineer the networks from measurements taken from its individual components. In the last few years, several methods for the reconstruction of regulatory networks and biochemical pathways from data have been proposed. These methods were reviewed for example in [1, 2].

J.C. Setubal and N.F. Almeida (Eds.): BSB 2013, LNBI 8213, pp. 148–159, 2013.

There are many different methods that can be applied to model Genetic Regulatory Networks (GRN). The most detailed and faithful model to represent GRNs are systems of Ordinary Differential Equations. They can accurately describe biophysical processes e.g. the intra-cellular processes of transcription factor binding, diffusion, and RNA degradation; see, for instance, [3, 4]. The detailed descriptions of the dynamics, as provided by ODEs, are essential to an exact understanding of GRNs but this comes with a price. The application of ODEs demands a huge amount of prior knowledge about the system under investigation. First, it is necessary to define how the components of the system are connected and second, all the parameters of the biochemical reactions have to be determined. Thus, albeit ODEs being one of the most accurate representation of GRNs its application is restricted by the need of substantial prior knowledge about the system they are representing.

Clustering methods are at the other extreme in representing GRNs. These methods have been extensively applied to the analysis of microarray gene expression data [2, 5]. Despite its low computational costs, clustering methods can only extract qualitative information about co-expression. These methods are not suitable to infer the detailed structure of the underlying biochemical signalling pathways.

Machine Learning methods provide a promising compromise between these two extremes. It allows interactions between nodes in the network to be represented in an abstract way and to infer these interactions from data in a systems context. Clearly the interactions represented in this way do not encompass the level of detail of the underlying pathways described by ODEs models. Nonetheless the Machine Learning methods are able to distinguish direct interactions from indirect interactions that are mediated by other nodes in the domain.

In this work we investigate the application of an Estimation of Distribution algorithm (EDA) in the task of inferring GRNs modelled as Bayesian Networks (BNs). We apply a score-based inference scheme where a score is assigned to a particular model (network structure) given some observed data. In order to evaluate the performance of our algorithm we apply it to three distinct source of data: (i) data generated from a multivariate Gaussian distribution, (ii) data generated with the GeneNetWeaver tool and (iii) real data from flow Cytometry experiments.

In section 2 the BNs, its score-based inference and EDAs are introduced. Section 3 presents a description of the utilized data sets, the set up of the simulations and the criteria we applied to evaluate the performance of the algorithm. In section 4 we present the results of the simulations and in last section, 5, we discuss the results leading to conclusions and directions for future work.

2 Methodology

In this work we use Bayesian Networks (BNs) as the model for the regulatory networks. BNs are probabilistic models and are specially suited to modelling relationships in noisy domains as is the case in biological systems and biological systems measurements. In order to search for BNs that can better explain the data that we have at our disposal we apply an score-based inference scheme. In this scheme an Estimation of Distribution algorithm is employed to the task of finding meaningful networks.

2.1 Bayesian Networks (BNs)

Bayesian Networks combine both probability theory and graph theory in one modelling scheme. Formally a graphical structure $\mathcal{M}$, a family of conditional probability distributions $\mathcal{F}$ and the parameters $\mathbf{q}$ fully specify a BN. The graph structure $\mathcal{M}$ of a BN is defined by a set of directed edges and a set of nodes. If we have a directed edge from node X_k to node X_j, then X_k is called *parent* of X_j. Random variables are represented by the nodes and conditional dependence relations are characterized by the edges. The essence of the interactions between nodes and the intensity of these interactions is represented by the family of conditional probability distributions $\mathcal{F}$ and their parameters $\mathbf{q}$. A BN is defined by a simple and unique rule for expanding the joint probability in terms of simpler conditional probabilities. This follows the local Markov property: *A node is conditionally independent of its non descendants given its parents.* Due to this property, the structure $\mathcal{M}$, of a BN has necessarily to be a directed acyclic graph (DAG), that is, a network without any directed cycles. Let $X_1, X_2, ..., X_N$ be a set of random variables represented by the nodes $i \in \{1, ..., N\}$ in the graph, define $\pi_i[\mathcal{M}]$ to be the parents of node X_i in graph $\mathcal{M}$, and let $X_{\pi_i[\mathcal{M}]}$ represent the set of random variables associated with $\pi_i[\mathcal{M}]$. Then we can write the expansion for the joint probability as $P(X_1, ..., X_N) = \prod_{i=1}^{N} P(X_i|X_{\pi_i[\mathcal{M}]})$.

2.2 Score-Based Approach Inference

In a score-based inference the aim is to devise a BN structure (DAG) from a given set of training data $\mathcal{D}$. The found DAG structure shall be the one that better explains the available data. In other words, if we define that $\mathbb{M}$ is the space of all models, the main objective is to find a model $\mathcal{M}^* \in \mathbb{M}$ that is most supported by the data $\mathcal{D}$, $\mathcal{M}^* = \text{argmax}_{\mathcal{M}} \{\mathcal{P}(\mathcal{M}|\mathcal{D})\}$. Having the best structure $\mathcal{M}^*$ and the data $\mathcal{D}$, we can now find the best parameters, $\mathbf{q} = \text{argmax}_{\mathbf{q}} \{P(\mathbf{q}|\mathcal{M}^*, \mathcal{D})\}$. If we apply Bayes' rule we get $P(\mathcal{M}|\mathcal{D}) \propto P(\mathcal{D}|\mathcal{M})P(\mathcal{M})$ where the marginal likelihood implies an integration over the whole parameter space:

$$P(\mathcal{D}|\mathcal{M}) = \int P(\mathcal{D}|\mathbf{q}, \mathcal{M})P(\mathbf{q}|\mathcal{M})d\mathbf{q} \tag{1}$$

The integral in (1), our score, is analytically tractable when the data is complete and the prior $P(\mathbf{q}|\mathcal{M})$ and the likelihood $P(\mathcal{D}|\mathbf{q}, \mathcal{M})$ satisfies certain regularity conditions [6, 7]. In this work we use the score proposed in [6] known as the Bayesian Gaussian likelihood equivalent (BGe) score. The BGe score assumes that the data comes from a multivariate Gaussian distribution. Following the Equation 1 we have a manner to assign a score to a graphical structure given a data set. Nonetheless, the search for high scoring structures is not trivial. Due to the fact that the number of structures increases super-exponentially with the number of nodes in the network it is impossible to list all the possible structures. Furthermore, when considering a sparse data set, $P(\mathcal{M}|\mathcal{D})$ is diffuse, meaning that it will not be properly represented by a single structure $\mathcal{M}^*$.

Hence, we apply an EDA to find high scoring structures. In order to have a collection of structures rather than a single result we run the algorithm 30 times and use the average of the results (DAGs) as our resulting network.

2.3   Estimation of Distribution Algorithm (EDA)

In evolutionary computation, the identification and preservation of important interactions among genes is called linkage learning. A recent survey [8] revises and summarizes existing linkage learning techniques for evolutionary algorithms.

The adoption of probabilistic models for selected individuals is a powerful approach for evolutionary computation. Probabilistic models based on high order statistics have been used by (EDAs) [9], resulting better effectiveness when searching for global optima for hard optimization problems [10].

An EDA solves a problem by building successive probabilistic models from the solutions in the population. New solutions (individuals) are sampled from that model. This avoids the adoption of the traditional crossover and mutation as in Genetic Algorithms, which operate on single individuals.

One way of classifying the EDAs is according to the employed probabilistic model [11]. The simpler order-1 EDAs such as Population-Based Incremental Learning (PBIL) [12] and Univariate Marginal Distribution Algorithm (UMDA) [13] adopt probabilistic models which assume independence among variables. The order-2 EDAs, e.g. Mutual Information Maximizing Input Clustering (MIMIC) [14] and Bivariate Marginal Distribution Algorithm (BMDA) [15], consider pairwise interactions among variables. The high order statistics EDAs takes in consideration potentially all the (in)dependence relationships among variables, see for instance Bayesian Optimization Algorithm (BOA) [16] and Estimation of Bayesian Network Algorithm (EBNA) [17].

Order-1 EDAs work poorly on problems with variable interactions. Studies on the dynamics of the evolutionary process of PBIL shows that, from the standpoint of dynamical systems, only the local optima of the search space are stable stationary points, so the algorithm is expected to converge to local optima [18]. The inability of order-1 EDAs on problems which present nonlinear interactions among variables prevents those algorithms from solving a broader class of problems.

Higher order statistics improve the chance of finding the global optimal solution, as shown in [19]. This leads to the class of higher order EDAs, which are based on learning the linkage among genes by inferring expressive probabilistic models based on searching for a factorization, which captures the dependencies among genes. Good results are reported for several problems in the literature whereas this class of EDAs imposes a high computational cost associated to the model induction stage. Finding a factorization can be a computationally expensive process and the resulting graph is often a suboptimal solution [20]. In general, the asymptotic time complexity of the search for a dependence structure in higher order EDAs dominates the overall complexity of the whole probabilistic model building [20].

In the present work we apply a novel approach proposed in [21]. The algorithm 1 is an EDA which employs clustering and probabilistic model learning. It also presents a new recombination operator, the cg-recombination.

Algorithm 1. φ-PBIL
___

//Initialization:
Generate an initial random population of size N_0.
Compute the fitness of the solutions.
//Learning:
while convergence criteria are not met **do**
    Compute a probability vector (PV) of binomial proportions for each cluster.
    randomly selects one PV, PV_a
    if utilize cg-recombination **then**
        randomly chooses another PV, PV_b
        Compute the matrix of information measures $W = (w_{i,j})$.
        Update PV_a guided by the matrix W
    end if
    Create a solution H accordingly with the chosen PV_a.
    compute the fitness of the new solution H
    if H is not worse than the worst in population **then**
        Delete this worst individual.
        Insert H in the population.
    end if
    Update clusters and matrixes
end while

The algorithm initializes by generating N_0 solutions thus defining the initial population. For each of these solutions a fitness is then calculated. At each iteration of the learning procedure a k-means algorithm is applied to create k population clusters where k is empirically defined. Distinctively from other EDA applications [22, 23] the present algorithm creates only one solution, H at each iteration. Therefore, only cluster centroids are updated avoiding the need of running the algorithm from the beginning. In order to generate a new solution there are two possibilities: (i) from a probability vector (PV) (obtained for each cluster) considering that the variables are independent; (ii) through the application of the cg-recombination operator which performs a combination among two PVs. The application of the cg-recombination operator is selected according to a set probability p_c. In the case where the cg-recombination is applied two PVs are randomly selected to generate the new individual. In this case $W_{i,j}$, the measure of how informative each parent cluster i is for each variable j, is calculated. The $W_{i,j}$ is defined as the difference between the entropy of the distribution of each variable j before and after observing a given cluster i. Its value guides the choice among the two selected PVs where the new individual is created from the most informative parts of the two selected PVs. For a comprehensive explanation of the algorithm the reader is directed to [21].

3   Simulations

3.1   Data Sets

In order to evaluate our proposed method we apply it to three distinct types of data. In increasing order of inference difficulty the data sets are: (i) data generated from a multivariate Gaussian distribution, (ii) data generated with the GeneNetWeaver tool and (iii) real data from flow Cytometry experiments. The difficult in this context is related with the agreement between the data generation model and the inference model.

Gaussian Multivariate Data: A clear and simple way of generating synthetic data from a given structure is to sample them from a linear Gaussian distribution. The random variable X_i denoting the expression of node i is distributed according to $X_i \sim N\left(\sum_k w_{ik} x_k, \sigma^2\right)$ where $N(.)$ denotes the Normal distribution, the sum extends over all parents of node i, and x_k represents the value of node k. The interaction strength between nodes X_i and X_k is $w_{ik} \neq 0$. If $w_{ik} = 0$ then node X_k is not a parent of node X_i. The value of σ^2 can be interpreted as being the dynamic noisy. Low values of σ^2 indicates a very deterministic data set, i.e., the value of the child node is completely determined by their parents values. Contrarily, high values of σ^2 indicates a noisy data set. This process is equivalent of sampling from a multivariate Gaussian distribution. It is important to note that in order to generate data for a given network structure it is necessary to topological sort the nodes first. This is fundamental to guarantee that the parent nodes have their values computed before their child nodes. To generate Gaussian data we set $w_{ik} = 1$ if the edge is present in the network and $w_{ik} = 0$ otherwise. We also set $\sigma^2 = 0.01$. It is interesting to observe that the Gaussian multivariate data is a perfect match for the scoring method, BGe, used in the present work. The data generated with this method will be referred in this work as Gaussian data.

GeneNetWeaver Data: In order to have more realistic simulated data we use the tool GeneNetWeaver (GNW) [25]. Data generated using GNW is obtained from a stochastic system of coupled differential equations (ODEs) with added noise. This type of data is more similar to real data as it presents non-linearities which are typical of real biologycal systems. However, because the data is simulated from a known structure, we are sure about what the answer of the inference algorithm shall be. In order to generate the data we used the GNW tool. We selected experiments to be "*multifactorial*" with "*add Gaussian noise*" and "*std dev = 0.005*". Data generated with this method from here onwards will be refereed as GNW.

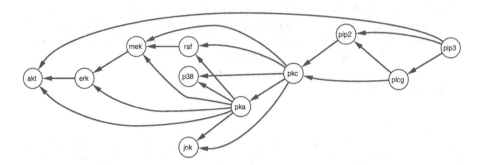

Fig. 1. Raf signalling pathway. The graph shows the currently accepted signalling network, taken from [24]. Nodes represent proteins, edges represent interactions, and arrows indicate the direction of signal transduction.

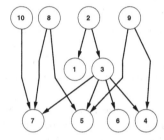

Fig. 2. Sub-network *Escherichia Coli*. The graph shows a sub-network extracted from *Escherichia Coli* network. This sub-network is part of the DREAM challenge 3 as presented in [25].

Real Flow-Cytometry Data: In [24] the authors used intracellular multi-colour flow cytometry experiments to measure the concentration levels of the 11 proteins that compose the network depicted in Fig. 1. This network is involved in regulating cellular proliferation in human immune system cells. The deregulation of this pathway can lead to carcinogenesis, and the pathway has therefore been extensively studied in the literature (e.g. [24, 26]). Because this network is widely studied, an accepted gold standard network obtained from various distinct studies is available and is presented in Fig.1. The data produced with this method is regarded as Real data in this work.

For each one of the three types of data, Gaussian, GNW and Real, we generated five data sets with 100 measurements (data points). The GNW and Real data sets were preprocessed before being analysed. We used quantile-normalisation to normalise each of the five data sets. That is, for each of the variables we replaced the 100 measured values by quantiles of the standard normal distribution $N(0, 1)$. More precisely, for each of the variables the j-th highest measured value was replaced by the $\left(\frac{j}{100}\right)$-quantile of the standard normal distribution, whereby the ranks of identical measured values were averaged.

For both types of simulated data, Gaussian and GNW, we obtained data sets from the structure presented in Fig. 2. The Real data sets are from the structure presented in Fig 1.

3.2   Simulation Set Up

In order to empirically validate and evaluate the application of the EDA in the inference of the GRNs modelled with BNs we applied it to the three types of data described in the previous section. For each type of data we apply the algorithm in each one of the five data sets.

Because the data is sparse and thus the information about variables relationships present in the data is diffuse it is very unlikely that a single resulting network will adequately represent the true network. Hence, we run the simulation for each data set 30 times and use the average of resulting networks as our final result.

In our EDA each solution to the problem is coded as an adjacency matrix representing a BN structure, $\mathcal{M}$. In each execution of the EDA the initial population is generated whereas each individual (solution) is subjected the following criteria: **(i)** each variable can have only 3 regulators (parents); **(ii)** the network is not allowed to have directed cycles and **(iii)** The zero (0) and one (1) entries in the adjacency matrix are sampled respectively with probability 65% and 35%.

The restriction in the number of regulators, known as fan-in restriction, is typical and has been applied e.g. in [27–29]. The BN is represented by a DAG and, therefore, directed cycles are not allowed. Also, because it is known that GRNs are sparsely connected we impose sparsity in the random generated networks by sampling the zero and one entries with different probabilities.

For each of the proposed solutions, $\mathcal{M}$, in the population, a fitness is calculated. The fitness for the network, $P(\mathcal{D}|\mathcal{M})$, is obtained as proposed in [6] and is known as the BGe score.

The result of an EDA execution is one adjacency matrix. This execution is repeated 30 times and the solution considered for each data set is the average of the 30 resulting adjacency matrices, $\mathcal{R}$.

3.3   Evaluation Criteria

The result of the EDA simulations is a collection of network structures represented in an adjacency matrix. From this collection of matrices we obtain one average matrix, $\mathcal{R}$, where each entry r_{ij} indicates the average occurrence of each edge in the collection of inferred networks. As a means to assess EDA's performance it is necessary to compare its result with some known network. We call this known network the true network, $\mathcal{T}$, where the entries, $t_{ij} \in \{0,1\}$, indicate the presence and the absence of the connection between nodes X_i and X_j. In order to compare our resulting network $\mathcal{R}$ with the true network $\mathcal{T}$, we transform it to an adjacency matrix, $\mathcal{A}_{\mathcal{R}}(\epsilon)$, by imposing a threshold ϵ. Each entry of the adjacency matrix, a_{ij}, is 1 if $r_{ij} \geq \epsilon$ and 0 otherwise.

Having these two matrices, $\mathcal{T}$ and $\mathcal{A_R}(\epsilon)$, we can classify each of the edges into categories. An edge can be classified as: true positive (TP), false positive (FP), true negative (TN) or false negative (FN). Table 1 shows a summary of how the edges are classified into these categories.

Table 1. Classification of edges. This table shows how an edge is classified according to the values in the true matrix (t_{ij}) and in the adjacency matrix (a_{ij}). An entry that is equal to zero means that the edge from node X_i to node X_j is absent, conversely an entry that is equal to one means that the edge is present.

t_{ij}	a_{ij}	Category
0	0	TN
0	1	FP
1	0	FN
1	1	TP

The receiver operator characteristics (ROC) curve is obtained by varying the threshold ϵ and plotting the relative number of TP edges against the relative number of FP edges for each of the thresholds. Ideally we would compare the whole ROC curves but this is impractical. Therefore, we use the area under the ROC curve (AUC) as it summarizes the results for all the thresholds. A perfect predictor would produce an AUC value of 1.00. Conversely, a random predictor would produce an AUC value around 0.50. In general, bigger area values represent better predictors.

4   Results and Discussion

In Fig. 3 the average AUC scores and standard deviations over five data sets for each of the three types of data are presented. As expected the inference of regulatory networks from Gaussian data presents the best results. This is mostly due to the fact that the model used for inference is an exact match for the model applied to generate the data, namely the multivariate Gaussian distribution. The realization of the inference in this scenario is very valuable as it permits to verify if the inference method produces meaningful results. In this work the high values of AUC obtained in this data set indicates that the proposed inference method presents a good performance regarding to the reconstruction of the regulatory network structures.

The inference of regulatory networks from the GNW data presents poorer results. In this case we are sure about the structure from which we generated data from but we know that the data generating process produces highly non-linear interactions and these are a mismatch to the inference model. This mismatch may be the cause of the poor results.

The worst results are obtained when inferring networks from the Real data. This is expected as in this case the generating process is a real biological system and, therefore, it is very unlikely that it follows a multivariate Gaussian

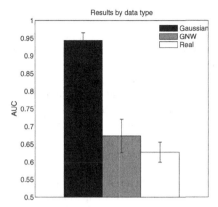

Fig. 3. Inference results. The graph shows the resulting average AUC scores for all three types of data, Gaussian (black), GNW (gray) and Real (white). The bars indicate the average AUC value over five data sets and the error bars present their respective standard deviation.

distribution. Also, the amount of noise in the biological process itself and in the measurement techniques are not known and poorly understood. Moreover, in this case, we have a very good indication of the true network, but the true network is not exactly known because it was manually curated from the available literature. Interestingly, these results are not very different from those obtained with the GNW data set. This indicates that the realistic simulated data is capable of generating close to real data sets.

5 Conclusion and Future Work

The present work shows that the proposed method of inferring GRNs using EDAs is feasible. The results are similar to a previous work [28], specially for the Real data. In [28] the authors have applied a traditional Markov Chain Monte Carlo (MCMC) Bayesian network inference algorithm to the same data set and the results obtained were very similar with the results obtained in the present work.

Although EDAs do not have a formal proof that they sample from the true posterior distribution, as the MCMC methods does, the similarity of their results is remarkable. This opens an interesting direction of research. In future work we plan to properly compare both inference methods applying rigours statistical tests to verify whether the results are significantly equal or not.

The idea of exploring the potential of EDAs in the inference of GRNs is mainly because they may be of great help for MCMC. The main problem of MCMC methods is that they suffer from problems of mixing and convergence, and are only guaranteed to sample from the posterior distribution if obeying strict conditions. This makes it difficult to propose alternative schemes of sampling in MCMC framework. On the other hand, EDAs do not have such restrictions and can be improved to run faster. Thus, we intend to empirically verify whether these results are relevant, testing it in various networks with different structures and sizes.

References

1. De Jong, H.: Modeling and simulation of genetic regulatory systems: A literature review. Journal of Computational Biology 9(1), 67–103 (2002)
2. D'haeseleer, P., Liang, S., Somogyi, R.: Genetic network inference: from co-expression clustering to reverse engineering. Bioinformatics 16(8), 707–726 (2000)
3. Chen, T., He, H.L., Church, G.M.: Modeling gene expression with differential equations. In: Pacific Symposium on Biocomputing, vol. 4, pp. 29–40 (1999)
4. Pokhilko, A., Fernández, A.P., Edwards, K.D., Southern, M.M., Halliday, K.J., Millar, A.J.: The clock gene circuit in Arabidopsis includes a repressilator with additional feedback loops. Molecular Systems Biology 8, 574 (2012)
5. Eisen, M.B., Spellman, P.T., Brown, P.O., Botstein, D.: Cluster analysis and display of genome-wide expression patterns. Proceedings of the National Academy of Sciences of the United States of America 95, 14863–14868 (1998)
6. Heckerman, D.: Learning Gaussian networks. Technical Report MSR-TR-94-10, Microsoft Research, Redmond, Washington (July 1994)
7. Heckerman, D.: A tutorial on learning with Bayesian networks. Technical Report MSR-TR-95-06, Microsoft Research, Redmond, Washington (1995)
8. Chen, Y.P., Yu, T.L., Sastry, K., Goldberg, D.E.: A survey of linkage learning techniques in genetic and evolutionary algorithms. Technical Report IlliGAL Report No. 2007014, University of Illinois at Urbana-Champaign (2007)
9. Larrañaga, P., Lozano, J.A.: Estimation of Distribution Algorithms: A New Tool for Evolutionary Computation. Kluwer (2002)
10. Emmendorfer, L.R., Pozo, A.: Effective linkage learning using low-order statistics and clustering. IEEE Transactions on Evolutionary Computation 13(6), 1233–1246 (2009)
11. Larrañaga, P., Karshenas, H., Bielza, C., Santana, R.: A review on evolutionary algorithms in Bayesian network learning and inference tasks. Inf. Sci. 233, 109–125 (2013)
12. Baluja, S., Caruana, R.: Removing the genetics from the standard genetic algorithm. In: International Conference on Machine Learning, pp. 38–46 (1995)
13. Mühlenbein, H., PaaB, G.: From Recombination of Genes to the Estimation of Distributions I. Binary Parameters. In: Ebeling, W., Rechenberg, I., Voigt, H.-M., Schwefel, H.-P. (eds.) PPSN 1996. LNCS, vol. 1141, pp. 178–187. Springer, Heidelberg (1996)
14. Bonet, J.S.D., Isbell, C.L., Viola, P.: Mimic: Finding optima by estimating probability densities. In: Jordan, M., Mozer, M., Perrone, M. (eds.) Advances in Neural Information Processing Systems, vol. 9, pp. 424–430. MIT Press, Cambridge (1997)
15. Mühlenbein, H.: The equation for response to selection and its use for prediction. Evol. Comput. 5(3), 303–346 (1997)
16. Pelikan, M., Goldberg, D.E., Cantu-Paz, E.: BOA: The Bayesian optimization algorithm. In: Proceedings of the 1999 Genetic and Evolutionary Computation Conference, pp. 525–532 (1999)
17. Etxeberria, R., Larrañaga, P.: Global optimization using Bayesian networks. In: Second Symposium on Artificial Intelligence (CIMAF 1999), pp. 332–339 (1999)
18. González, C., Lozano, J.A., Larrañaga, P.: Analyzing the PBIL algorithm by means of discrete dynamical systems. Complex Systems 4, 465–479 (2000)
19. Zhang, Q.: On stability of fixed points of limit models of univariate marginal distribution algorithm and factorized distribution algorithm. IEEE Transactions on Evolutionary Computation 8(1), 80–93 (2004)

20. Pelikan, M., Saltry, K., Goldberg, D.E.: Sporadic model building for efficiency enhancement of hBOA. Genetic Programming and Evolvable Machines (2008)
21. Emmendorfer, L.R., Pozo, A.T.R.: An incremental approach for niching and building block detection via clustering. In: Proceedings of the Seventh International Conference on Intelligent Systems Design and Applications, ISDA 2007, pp. 303–308. IEEE Computer Society, Washington, DC (2007)
22. Baluja, S.: Population-based incremental learning. Technical Report CMU-CS-94-163, Computer Science Dept., Carnegie Mellon University (1994)
23. Georges, R., Harik, F.G.L., Goldberg, D.E.: The compact genetic algorithm. IEEE Trans. Evolutionary Computation 3(4), 287–297 (1999)
24. Sachs, K., Perez, O., Pe'er, D., Lauffenburger, D.A., Nolan, G.P.: Causal protein-signaling networks derived from multiparameter single-cell data. Science 308(5721), 523–529 (2005)
25. Schaffter, T., Marbach, D., Floreano, D.: GeneNetWeaver: In silico benchmark generation and performance profiling of network inference methods. Bioinformatics 27(16), 2263–2270 (2011)
26. Dougherty, M.K., Müller, J., Ritt, D.A., Zhou, M., Zhou, X.Z., Copeland, T.D., Conrads, T.P., Veenstra, T.D., Lu, K.P., Morrison, D.K.: Regulation of Raf-1 by direct feedback phosphorylation. Molecular Cell 17, 215–224 (2005)
27. Friedman, N., Linial, M., Nachman, I., Pe'er, D.: Using Bayesian networks to analyze expression data. Journal of Computational Biology 7, 601–620 (2000)
28. Werhli, A.V., Grzegorczyk, M., Husmeier, D.: Comparative evaluation of reverse engineering gene regulatory networks with relevance networks, graphical Gaussian models and Bayesian networks. Bioinformatics 22(20), 2523–2531 (2006)
29. Husmeier, D.: Sensitivity and specificity of inferring genetic regulatory interactions from microarray experiments with dynamic Bayesian networks. Bioinformatics 19, 2271–2282 (2003)

A Network-Based Meta-analysis Strategy
for the Selection of Potential Gene Modules
in Type 2 Diabetes

Ronnie Alves[1], Marcus Mendes[2], and Diego Bonnato[2]

[1] Vale Institute of Technology, Belém, Brazil
ronnie.alves@vale.com
[2] Federal University of Rio Grande do Sul, Porto Alegre, Brazil
diego@cbiot.ufrgs.br, cla_atm_milo@hotmail.com

Abstract. We propose an integrative network-based meta-analysis strategy to enable the selection of potential gene markers for one of the most
prevalent diseases worldwide, Type 2 diabetes (T2D), formally known as
the non-insulin dependent diabetes mellitus. Comprehensive elucidation
of the genes regulated through this disorder and their wiring will provide
a more complete understanding of the overall gene network topology and
their role in disease progression and treatment. The proposed strategy
was able to find conservative gene modules which play interesting role in
T2D, pointing to gene markers such as NR3C1, ADIPOR1 and CDC123.
Network-based meta-analysis by enumerating conserved gene modules
pave a practical approach to the identification of candidate gene markers
across several related transcriptomic studies. The NEMESIS R pipeline
for network-based meta-analysis is also provided.

Keywords: gene co-expression network analysis, candidate gene markers, type 2 diabetes, meta-analysis, system biology.

1 Introduction

Type 2 diabetes (T2D), formally known as the non-insulin dependent diabetes
mellitus, is the most common type of diabetes. It has been taking the quality of
an endemic disease, affecting more than 170 millions of people around the world.
In fact, there is tragic trend that by the end of 2030 more than 300 millions of
people will develop T2D.

In T2D, either the body does not produce enough insulin or the cells do
not respond to the insulin. The effects of insulin, insulin deficiency and insulin
resistance vary according to the physiological function of the organs and tissues
concerned, and their dependence on insulin for metabolic processes. Those tissues
defined as insulin dependent, based on intracellular glucose transport, are mainly
adipose tissue and muscle. Although insulin resistance occurs in most obese
individuals, diabetes is usually forestalled through compensation with increased
insulin. This increase in insulin occurs through an expansion of beta-cell mass

J.C. Setubal and N.F. Almeida (Eds.): BSB 2013, LNBI 8213, pp. 160–169, 2013.
© Springer International Publishing Switzerland 2013

and/or increased insulin secretion by individual beta-cells. Failure to compensate for insulin resistance leads to T2D [1–3].

T2D is a multifactorial disease caused by both oligo- and polygenic genetic factors as well as non-genetic factors that result from a lack of balance between the energy intake and output and other life style related factors. There is a plenty of data related to the genetics of T2D. Though, many genes and gene products as well as their interactions with the environment at the molecular, cellular, tissue, and the whole organism levels are still unknown. Understanding of diabetes pathogenesis is critical to the development of new strategies for effective prevention and treatment of this disease [4–6].

With the advance of high-throughput technologies, it is now possible to get deep insight into the orchestration of the complex biological functions either activated or not during disease development. Genome-wide association studies (GWASs) have discovered association of several loci with Type 2 diabetes. In such studies, gene expression profiles are evaluated from a set of several associated studies where genes presenting stable modulation patterns across these studies could be potential markers. However in most cases they do not take into account the network interaction of the genes involved in the correlated T2D pathways [7].

In the present work we propose a network-based meta-analysis strategy for the selection of potential gene regulatory modules in T2D. The basis to retrieve such potential gene modules relies on the proper inference of weighted gene co-expression networks from associated T2D transcriptomic studies. Next, gene modules identified among these studies are evaluated for functional enrichment of the conserved gene modules (*consensus cliques*).

The main contributions of the proposed strategy are:

- An empirical evaluation of weighted gene co-expression network on T2D studies;
- A new method for the localization of functional consensus gene modules through frequent pattern mining;
- The NEMESIS R pipeline for the selection of potential gene regulatory modules, revealing biological functions correlated to T2D pathogenesis.

The remainder of this paper is organized as follows. In Section 2 we present the transcriptomic data sets as well as the methods used by the proposed strategy. Next, in Section 3 we present the pipeline developed and the main results obtained. Conclusions and future work are provided in Section 4.

2   Materials and Methods

2.1   Transcriptomic Data Sets

We carefully selected four transcriptomic studies from the Gene Expression Omnibus (GEO) which were properly elaborated to measure T2D progression and treatment. In the first dataset (GSE12389) diverse roles were demonstrated regarding interferon-gamma (IFN-γ) in the induction and regulation of immune-mediated inflammation using a transfer model of autoimmune diabetes [8]. The second dataset (GSE2253) investigated the molecular mechanism by which extracellular hIAPP mediates pancreatic beta-cell apoptosis [9]. It is known that extracellular hIAPP oligomers are toxic to pancreatic beta-cells and associated with apoptosis. The third study (GSE12639) highlights genetic regulatory mechanisms in the remote zone of left ventricular (LV) free wall in order to partly explain the more frequent progression to heart failure after acute myocardial infarction (AMI) in diabetic rats [10]. And finally, the fourth dataset (GSE13270) studied Type 2 diabetes progression and the development of insulin resistance in two animal models with and without a high fat diet superimposed on these models [11].

2.2   Preprocessing Affymetrix Data

The Affymetrix expression data (CEL files) were preprocessed using the Robust Multi-array Average (RMA) normalization approach through the rma() function in the *affy* R package. RMA employs quantile normalization and smooths technical sources of variability across samples. Next, only expressed genes were selected for further analysis. A gene (probe) was considered expressed if it was called (P)resent or (M)arginal in at least 75% of all samples in a given dataset. Present and Marginal calls were determined by the mas5calls() function in the *affy* R package. A detection call answers the question: *Is the transcript of a particular gene Present or Absent?* In this context, absent means that the expression level is below the threshold of detection. That is, the expression level is probably not different from zero. In the case of an uncertainty, we can get a marginal call. It is important to note that some probe-sets are more variable than others, and the minimal expression level provably different from zero may range from a small value to very large value (for a noisy probe-set). The advantage of asking the question in this way without actual expression values is that the results are easy to filter and to interpret. For example, we may only want to look at genes whose transcripts are detectable in a particular experiment. Given that co-expressed gene networks are created based on correlation metrics over the study, such fllter strategy allow us to remove potential inconsistencies in the Chip. Thus, it was used a cutoff of 75% with only one particular exception for the GSE12389 study (50%). Table 1 presents the results of the preprocessing step over all selected studies.

2.3   Calculating Candidate Gene Modules

The network topology of each study was calculated through the application of the WGCNA R package [12], and various soft-thresholding powers were properly applied to find a good fitness of the scale-free topology. The soft-thresholding strategy, adopted by WGCNA, keeps all possible links and raises the original

Table 1. Data preprocessing of Affymetrix datasets

Study	Samples	Genes before	Genes after	Affy.Chip
GSE12389	8	45101	5316	Mouse430_2
GSE2253	20	22690	11686	Mouse430A
GSE12639	12	31099	14998	Rat230_2
GSE13270	101	31099	13935	Rat230_2

coexpression values to a power "beta" so that the high correlations are emphasized at the expense of low correlations. An example of the scale-free topology calculated for the GSE13270 study is presented in Figure 1.

Once the network has been constructed, module inference is the next step. Modules are defined as clusters of densely interconnected genes. WGCNA detects gene modules using unsupervised clustering, i.e. without the use of a priori defined gene sets. In fact, modules are calculated based on a topological overlap measure [13] that has been applied successfully in several applications. The user has a choice of several module detection methods. The default method is hierarchical clustering using the standard R function hclust, branches of the hierarchical clustering dendrogram correspond to modules and can be identified using one of a number of available branch cutting methods, for example the constant-height cut or two Dynamic Branch Cut methods. One drawback of hierarchical clustering is that it can be difficult to determine how many (if any) clusters are present in the data set. Although the height and shape parameters of the Dynamic Tree Cut method provide improved exibility for branch cutting and module detection, it remains an open research question how to choose optimal cutting parameters or how to estimate the number of clusters in the data set [12].

The final power threshold p calculated to each data set was defined as follows: GSE12389 ($p=18$), GSE2253 ($p=9$), GSE12639 ($p=12$) and GSE13270 ($p=12$). Next, modules for each study were extracted, and only significant intramodular (hub) genes at each module were selected for further analysis. We use the intramodular connectivity measure to define the most highly connected intramodular hub gene as the module representative. In fact, intramodular hub genes are highly correlated with the module eigengene. A gene in a module is considered significant if it has a strong p-value (< 0.001) membership, i.e., correlation between module eigengenes and expression values in the Chip (Figure 2). In Table 2 the effects of the calculation of the candidate genes are presented to each corresponding study.

2.4 Functional Enrichment Analysis

We determined the specific biological processes relevant for each candidate gene module by calculating GO terms and pathway enrichment. Furthermore, each module is also a generalized clique [13]. We obtained significant (p-value<0.05)

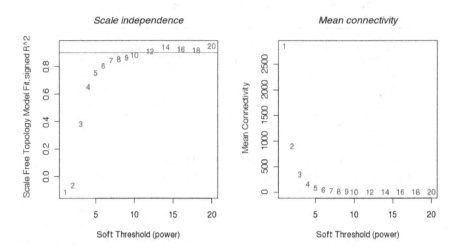

Fig. 1. Analysis of network topology for various soft-thresholding powers for the GSE13270 study. The left panel shows the scale-index (y-axis) as a function of the soft-thresholding power (x-axis). The right panel displays the mean connectivity degree (y-axis) as a function of the soft-thresholding power (x-axis).

Table 2. Data preprocessing of Affymetrix datasets

Study	Modules	Genes before	Genes after	Affy.Chip
GSE12389	12	5316	1566	Mouse430_2
GSE2253	17	11686	5570	Mouse430A
GSE12639	34	14998	5660	Rat230_2
GSE13270	16	13935	9836	Rat230_2

GO and pathway enrichment for all modules. The respective *Entrez* gene identification was obtained through the *biomaRt* R package. Next, we make use of the *GOstats* R package as well as the related *Affymetrix Chip Expression Set* annotation data to each associated organism.

2.5   Finding Consensus Modules

Consensus modules were detected by exploring the retrieved functional annotations and evaluating the co-occurrence of these annotations across the related T2D studies. Thus, whether a significant annotation is shared among distinct modules, across several networks, such observation could be seen as good indication of consensus. The intuition of exploring co-occuring gene sets has been explored broadly by the data mining community in several gene association studies [14]. However, as far as we are concerned, there is no direct reference of its utilization in network-based meta-analysis. In this work, we say that we

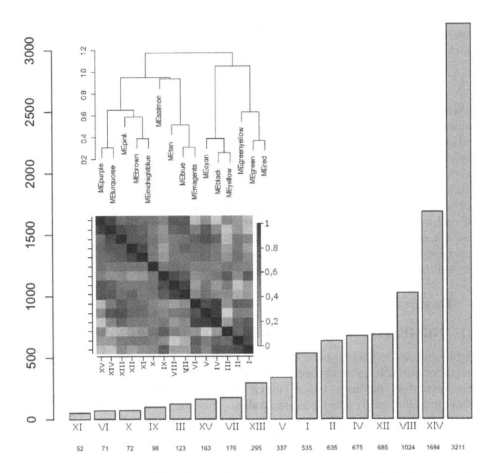

Fig. 2. Heatmap plot of the adjacencies in the eigengene network including the trait weight build on the GSE13270 study. Each row and column in the heatmap corresponds to one module eigengene (labeled by color) or weight. In the upper dendogram (heatmap), white color represents low adjacency (negative correlation), while black represents high adjacency (positive correlation). Squares of black color along the diagonal are the gene modules (Bar plots in the bottom). Genes that were not assigned to any modules were assigned to the largest Bar plot on the right.

have a consensus module when its associated annotation is shared across different studies. The consensus significance is measured by metrics like *support* and *confidence* of the co-occurring annotations. Therefore, before exploring such patterns we have to introduce two concepts called *Transactions* and *Item set* . Each transcriptomic study is related to one *transaction_id* and it is composed by several gene modules (i.e, either GO or KEGG annotations). An *item set* is an annotation (or a set of annotation) that appears in more that one study. Thus, if a particular annotation has a support of 75%, it does mean that this functional behavior is observed in three out of four related studies (see Transcriptomic data sets). By using such consensus strategy we avoid the hard task to conciliate all different gene names in all distinct Affymetrix platforms and organisms, focusing on the search of functional gene modules closely related to T2D pathology.

2.6 Selection of Potential T2D Genes

Potential genes are those ones that are significantly covered by the enriched consensus modules. Since only the most conserved annotations are selected for further analysis, it was necessary to devise a reverse engineering approach for the identification of the consensus genes. Thus, we first selected all associated genes to the most relevant *GO terms* with its *EntrezGene* information. Further, this gene set was matched with the gene list obtained by the consensus analysis. For the GO information we used the functional annotation *GO.db* database, and for each organism its associated species annotation database. For instance to the *mus musculus* we selected the *org.Mm.eg.db*. We have also used the *Phenopedia* database [15] to evaluate the correspondence of the selected potential genes with the well-known T2D (*human*) genes induced by this database.

3 Results and Discussion

3.1 NEMESIS: The *NE*twork-Based *ME*ta-analy*SIS* Pipeline

Meta-analysis has been applied broadly in several disease studies to improve the search for potential gene markers. Network-based strategies highlight important gene regulatory modules, but it cannot ensure that such module(s) could be conserved along with other related studies. Therefore, one pontential alternative, as proposed here, is the combination (or meta-analysis) of several gene co-expression networks to pulling out gene modules providing relevant functional association to the disease phenotype. Once having all selected transcriptomic data sets for analysis the first step is the proper normalization procedure. Next, gene modules are enumerated by exploring network functions available in the WGCNA *R* package. The following task is then the functional enrichment analysis to retrieve significant gene annotations related to network modules. After getting these annotations, it is possible to search for the conserved biological functions by exploring frequent patterns on the annotated modules. Finally, candidate gene markers are evaluated through the mapping of the biological functions associated with pathology under investigation. The proposed pipeline is summarized in Figure 3. The *R* scripts and additional material are free available online at https://sites.google.com/site/alvesrco/nemesis.

3.2 Potential T2D Gene Markers

The presented strategy was able to enumerate potential gene markers correlated to T2D. For instance, the NR3C1 gene, being also a well-known product of a transcription factor highly associated to T2D.

Next, we highlight the main candidate genes retrieved by the NEMESIS pipeline applied on T2D transcriptomic data sets:

- The gene ADIPOR1 encodes a protein which acts as a receptor for adiponectin, a hormone secreted by adipocytes which regulates fatty acid catabolism and glucose levels. Binding of adiponectin to the encoded protein results in activation of an AMP-activated kinase signaling pathway which affects levels of fatty acid oxidation and insulin sensitivity. Patients who developed T2D present a low activity of this gene when compared with normal ones [16];
- The gene CDC123 encodes proteins highly associated to the production of insulin. Variations of this gene are also related to a low production of the hormone [17];
- The gene SERPINE1 encodes a member of the serine proteinase inhibitor (serpin) superfamily. This member is the principal inhibitor of tissue plasminogen activator (tPA) and urokinase (uPA), and hence is an inhibitor of fibrinolysis. Comparative proteomic profiling of plasma from individuals with either diabetes or obesity and individuals with both obesity and diabetes revealed SERPINE 1 as a possible candidate protein of interest, which might be a link between obesity and diabetes [3].

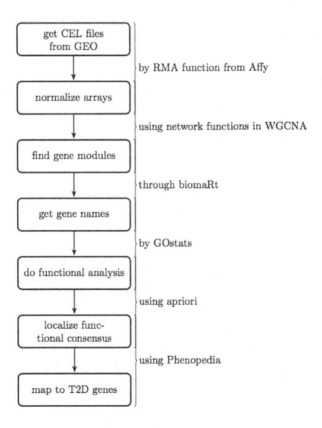

Fig. 3. The NEMESIS R pipeline to explore network-based meta-analysis on transcriptomic data

4 Conclusions

We have introduced a network-based meta-analysis approach to discover potential gene modules, biologically associated, to type 2 diabetes pathology. Though, we also envisage its application on other complex diseases, such as those ones with large collection of transcriptomic data available in the Phenopedia database. The NEMESIS *R* pipeline developed could be easily extended to explore gene markers on other diseases.

In the present study four transcriptomic studies were selected for the experimental analysis. Despite the good results, it would be interesting to explore the pipeline with more T2D data sets to increase significance of the discovered patterns. However, the more data we use the more preprocessing efforts are necessary in order to reduce data dimensionality, smoothing the creation of the associated gene co-expression networks. And consequently, the search for consensus gene modules.

There are plenty of open challenges with the network-based meta-analysis. We list the following directions to pursue in near future: i) a more compact and semi-automatic way to identify gene network modules (*cliques*), ii) an optimization procedure to calculate the thresholds for the most relevant annotation across studies, and iii) extend the pipeline to deal with RNA-Seq data.

Acknowledgements. We would like to thank reviewers for helpful suggestions and criticisms that have contributed to improve substantially the article. We also thanks Wallace Lira for the preparation of Figure 2. This work is partially supported by the Brazilian National Research Council (CNPq – *Universal calls*) under the BIOFLOWS project [475620/2012-7].

References

1. Liu, M., Liberzon, A., Kong, S.W., Lai, W.R., Park, P.J., Kohane, I.S., Kasif, S.: Network-based analysis of affected biological processes in type 2 diabetes models. PLoS Genet. 3(6), e96 (2007)
2. Stumvoll, M., Goldstein, B.J., van Haeften, T.W.: Type 2 diabetes: principles of pathogenesis and therapy. Lancet 365(9467), 1333–1346 (2005)
3. Kaur, P., Reis, M.D., Couchman, G.R., Forjuoh, S.N., Greene, J.F., Asea, A.: Serpine 1 links obesity and diabetes: A pilot study. J. Proteomics Bioinform. 3(6), 191–199 (2010)
4. Goh, K.I., Cusick, M.E., Valle, D., Childs, B., Vidal, M., Barabási, A.L.: The human disease network. Proc. Natl. Acad. Sci. U. S. A. 104(21), 8685–8690 (2007)
5. Keller, M.P., Choi, Y., Wang, P., Davis, D.B., Rabaglia, M.E., Oler, A.T., Stapleton, D.S., Argmann, C., Schueler, K.L., Edwards, S., Steinberg, H.A., Chaibub Neto, E., Kleinhanz, R., Turner, S., Hellerstein, M.K., Schadt, E.E., Yandell, B.S., Kendziorski, C., Attie, A.D.: A gene expression network model of type 2 diabetes links cell cycle regulation in islets with diabetes susceptibility. Genome Res. 18(5), 706–716 (2008)
6. Park, K.S.: Prevention of type 2 diabetes mellitus from the viewpoint of genetics. Diabetes Res. Clin. Pract. 66(suppl. 1), S33–S35 (2004)

7. Jain, P., Vig, S., Datta, M., Jindel, D., Mathur, A.K., Mathur, S.K., Sharma, A.: Systems biology approach reveals genome to phenome correlation in type 2 diabetes. PLoS One 8(1), e53522 (2013)
8. Calderon, B., Suri, A., Pan, X.O., Mills, J.C., Unanue, E.R.: Ifn-gamma-dependent regulatory circuits in immune inflammation highlighted in diabetes. J. Immunol. 181(10), 6964–6974 (2008)
9. Casas, S., Gomis, R., Gribble, F.M., Altirriba, J., Knuutila, S., Novials, A.: Impairment of the ubiquitin-proteasome pathway is a downstream endoplasmic reticulum stress response induced by extracellular human islet amyloid polypeptide and contributes to pancreatic beta-cell apoptosis. Diabetes 56(9), 2284–2294 (2007)
10. Song, G.Y., Wu, Y.J., Yang, Y.J., Li, J.J., Zhang, H.L., Pei, H.J., Zhao, Z.Y., Zeng, Z.H., Hui, R.T.: The accelerated post-infarction progression of cardiac remodelling is associated with genetic changes in an untreated streptozotocin-induced diabetic rat model. Eur. J. Heart Fail 11(10), 911–921 (2009)
11. Almon, R.R., DuBois, D.C., Lai, W., Xue, B., Nie, J., Jusko, W.J.: Gene expression analysis of hepatic roles in cause and development of diabetes in goto-kakizaki rats. J. Endocrinol. 200(3), 331–346 (2009)
12. Langfelder, P., Horvath, S.: Wgcna: an r package for weighted correlation network analysis. BMC Bioinformatics 9, 559 (2008)
13. Zhang, B., Horvath, S.: A general framework for weighted gene co-expression network analysis. Stat. Appl. Genet. Mol. Biol. 4, Article17 (2005)
14. Alves, R., Rodriguez-Baena, D.S., Aguilar-Ruiz, J.S.: Gene association analysis: a survey of frequent pattern mining from gene expression data. Brief. Bioinform. 11(2), 210–224 (2010)
15. Yu, W., Clyne, M., Khoury, M.J., Gwinn, M.: Phenopedia and genopedia: disease-centered and gene-centered views of the evolving knowledge of human genetic associations. Bioinformatics 26(1), 145–146 (2010)
16. Tomas, E., Tsao, T.S., Saha, A.K., Murrey, H.E., Zhang, C.C., Itani, S.I., Lodish, H.F., Ruderman, N.B.: Enhanced muscle fat oxidation and glucose transport by acrp30 globular domain: acetyl-coa carboxylase inhibition and amp-activated protein kinase activation. Proc. Natl. Acad. Sci. U. S. A. 99(25), 16309–16313 (2002)
17. Grarup, N., Andersen, G., Krarup, N.T., Albrechtsen, A., Schmitz, O., Jørgensen, T., Borch-Johnsen, K., Hansen, T., Pedersen, O.: Association testing of novel type 2 diabetes risk alleles in the jazf1, cdc123/camk1d, tspan8, thada, adamts9, and notch2 loci with insulin release, insulin sensitivity, and obesity in a population-based sample of 4,516 glucose-tolerant middle-aged danes. Diabetes 57(9), 2534–2540 (2008)

RAIDER: Rapid Ab Initio Detection of Elementary Repeats

Nathaniel Figueroa[1], Xiaolin Liu[2], Jiajun Wang[1], and John Karro[1,3,4]

[1] Department of Computer Science and Software Engineering
[2] Center for Molecular and Structural Biology
[3] Department of Microbiology
[4] Department of Statistics at Miami University, Oxford, OHIO, USA
karroje@miamiOH.edu

Abstract. Here we present RAIDER, a tool for the *de novo* identification of elementary repeats. The problem of searching for genomic repeats without reference to a compiled profile library is important in the annotation of new genomes and the discovery of new repeat classes. Several tools have attempted to address the problem, but generally suffer either an inability to run at the whole-genome scale or loss of sensitivity due to sequence variation between repeat copies. To address this, Zheng and Lonardi define *elementary repeats*: building blocks that can be assembled into a repeat library, but allow for the filtering of spurious fragments. However, their tool was too slow for use on large input, and subsequent attempts to improve efficiency have been unable to deal with the expected variation between repeat instances. RAIDER addresses both these problems, implementing a novel algorithm for elementary repeat detection and incorporating the spaced seed strategy of Pattern-Hunter to allow for copy variation. Able to process the human genome in under 6.4 hours, initial results indicate a coverage rate comparable to or better than that achieved by competing *de novo* search tool when paired with the library-based RepeatMasker.

1 Introduction

The identification of genomic repeats, sequences occurring multiple times within a genome, is a vital step in the study of genome structure. Repeat elements compose a substantial portion of most eukaryotic genomes (upwards of 45% in many mammalian genomes [13]). They are a source of genetic disorders, a mechanism for evolutionary change, and a window into genomic history that allows for the estimation of genomic substitution rates and the annotation of past evolutionary events [4,7,11,21].

Repeat discovery is currently best accomplished through tools that rely on a database of already annotated repeats, the most popular of which is Repeat-Masker [1]. This library-based approach allows for the identification of new members of known families, but it is incapable of discovering novel repeats. Hence the need for effective *de novo* search methods – algorithms that can identify

J.C. Setubal and N.F. Almeida (Eds.): BSB 2013, LNBI 8213, pp. 170–180, 2013.

repeat sequences throughout the genome without reliance on previously collected information.

There are significant obstacles to *de novo* repeat identification. The quadratic runtime of self-alignment algorithms make that approach infeasible for use on whole-genome input, and sequence variation (introduced by copy error and the effects of molecular evolution) stymies exact string matching algorithms. Transposable Elements (TEs), sequences with the ability to copy themselves to new locations, introduce a third complication when they enable hitchhiking: the transposition of a neighboring sequence fragment along with the TE which is easily miss-annotated as part of the element [4]. A final complicating factor is the mosaic pattern of subrepeats frequently underlying repeats, making it difficult to differentiate entire repeats from the basic elements making up their composition [17,24].

A number of computational approaches to the *de novo* repeat discovery problem are contrasted in Saha et al. [19,20]. But a method that is both practical on a whole-genome scale and able to account for the sequence variations has proven elusive. Tools based on self-alignment (e.g. RECON [3], PILER [5]) can account for sequence changes but do not scale well to whole-genome use. Tools that rely on the identification of fixed-length words (e.g. REPuter [12], ReAS [15], RepeatScout [18]) are able to improve runtime through the use of suffix-based tools, but are largely unable to handle sequence variation. A third approach is illustrated in RepeatGluer [17,24], which decomposes the repeats into an underlying mosaic pattern through a *De Bruijn* graph variation [22], but also does not scale well to larger sequences.

Another line of approach is the identification of *elementary repeats*, defined by Zheng and Lonardi [23] as the "building blocks" of repeat sequences – similar to the mosaic patterns addressed by RepeatGluer [17]. With this definition the author proposed an identification algorithm, but the algorithm's runtime was quadratic in the size of the query sequence, and thus not practical for whole-genome use. This was improved upon both by He [8] and Huo *et al.* [9], each employing suffix tree variations. However, such structures are geared towards searching for exact copies of substrings – hence their sensitivity suffers significantly in the presence of sequence variation between the repeats.

The tool RAIDER (**R**apid **A**b **I**nitio **D**etection of **E**lementary **R**epeats) identifies elemental repeats in linear time while allowing for sequence variation through the use of the PatternHunter spaced seed strategy [14,16]. With RAIDER we are able to process the Human Genome sequence in less than 6.4 hours on a single processor, and can prove RAIDER will identify all elementary repeats in the absence of sequence variation, and we can demonstrate the effectiveness of the PatternHunter augmentation to the basic algorithm to account for the presence of variation. RAIDER is open source, and can be downloaded from `http://handouts.cec.miamiOH.edu/karroje/RAIDER/` or by contacting the corresponding author.

2   Background

For purposes of this discussion we will focus on the detection of transposable elements (TEs): elements inserted into the genome by the repeated copying of some ancestral sequence and its descendants. Such elements naturally divide into families of homologous descendants distributed throughout the genome, leaving us with the challenge of identifying modern instances and grouping them into families without knowledge of the family progenitor sequences. At first glance, the natural approach is a seed-and-extend technique based on the identification of repeated l-mers (words of length l). But several factors compromise this approach. Sequence divergence resulting from molecular evolution will reduce the sensitivity of methods reliant on exact string matches, while the truncation of some repeat copies and artificial extension of others can result in spurious filtering of whole subsets of a repeat family.

To address these problems, Zheng and Lonardi define the concept of *elementary repeats* [23]: subsequences that serve as building blocks of the repeats, similar to the mosaic pattern identified by RepeatGluer and its successor [17,24]. Specifically, Zheng and Lonardi define elementary repeats as subsequences that meet minimum criteria for length and frequency, but that are not subsequences of other elementary repeats. Formally:

Definition 1. *(Zheng and Lonardi [23]) For a genomic sequence G and fixed values l and f: a sequence s, $|s| \geq l$, is an* **elementary repeat** *if: (1) All occurrences of s are maximally identical pairs; (2) s occurs at least f times, and (3) every substring of s of length $\geq l$ must occur exactly the same number of times in G as does s.*

Simply put: s is an elementary repeat if it has sufficient length, occurs with sufficient frequency, has no sufficiently long substrings that occur outside of s, and is not a subsequence of any sequence meeting this definition.

In searching for a linear algorithm able to account for sequence variation with repeat families, we turn to the spaced seed strategy of the PatternHunter homology-based search tool [14,16]. PatternHunter augments the BLAST *high scoring pair (HSP)* search step [2], replacing the requirement of exact string matches with a search for matches conforming to a *spaced seed*. Given a query string q and a target string t, BLAST will flag the pair for further investigation if q and t share a substring of some fixed length. PatternHunter instead requires they share matches dictated by a seed pattern – a binary string where 1s indicate the positions of required matches. In conjunction with BLAST, the use of spaced seeds significantly increases the sensitivity of the high-scoring pairs search with negligible effect on specificity. Here we incorporate spaced seeds into the definition of an elementary repeat, allowing for inexact matches. That is, we classify subsequences as belonging to the same elementary repeat family if they are a match according to some specified spaced seed (as opposed to being identical) and otherwise meet the conditions of the Zheng and Lonardi definition.

3   Methods

3.1   Algorithm Background

For ease of explanation, we will first assume there have been no base substitutions within our basic elements; this assumption will be relaxed later on. The method is built around a scan of the genome in which l-mer counts are tracked in a hash-table. During this scan, frequently occurring l-mers are expanded into maximal recurring patterns, then decomposed into *potential* elementary repeats. Any of these patterns surviving to the end of the scan are elementary repeats. (A preliminary scan to filter out low-frequency l-mers can be used to reduce memory requirements when working with genome-scale sequences, but does not change the steps of the main algorithm.) When encountering a new member of a potential family we need only compare it to the last member to identify any needed decomposition, allowing us to complete the scan in a runtime linear in the size of the genome. With our simplifying assumption, the algorithm is guaranteed to identify all instances of the elementary repeats.

We first need to define the "merge-by-overlap" operator: if x and y are strings, $x \circ y$ denotes the string created by merging x and y at the longest substring that is a suffix of x and a prefix of y. (e.g. $AACC \circ CCGG = AACCGG$). Note that, for a fixed value l any string s of length $\geq l$ can be formed by the merging of $|s| - l$ l-mers (that is: $s = x_0 \circ x_1 \circ \cdots \circ x_{|s|-l}$, where x_i is the l-mer starting exactly i bases after s in G). Given this, we now restate of the definition of elementary repeats, highlighting their connection to l-mers.

Definition 2. *For a genomic sequence G and fixed values l and f: a sequence s, $|s| \geq l$, is an elementary repeat if we can write $s = x_0 \circ \cdots \circ x_{|s|-l}$ $(|x_i| = l)$ such that: (1) there are at least f copies of s in G; (2) $x_i \neq x_j$ for $i \neq j$; (3) every occurrence of x_i in G is a substring of an occurrence of s in G; (4) s is maximal with respect to properties (1)-(3).*

Lemma 1. *The two definitions of repeat elements are equivalent.*

The proof is straightforward, once it is noticed that condition (3) of the original definition implies (2) and (3) of the new definition. (The reverse follows from the following lemma.)

Following from this definition is a lemma that will be useful in discussion of the algorithm:

Lemma 2. *Let x and y be maximal identical substrings of size $\geq l$ at two different locations of the genome. Neither can be a proper substring of an elementary repeat.*

Proof. Suppose w.l.o.g. that x is a proper substring of an elementary repeat e. y cannot be covered by a string identical to e, since x and y are maximal identical strings and $|e| > |x| = |y|$. But then e contains at least one l-mer (any l-mer in x) that has a copy (in y) that is not contained in a copy of e – hence violating condition (3) of our definition. □

It is this lemma that serves as the basis for our algorithm. When we find two maximal identical substrings, from the lemma we know that they might define an elementary repeat, or that they might contain an elementary repeat. But we need waste no more computation on the possibility that they (or any l-mer contained within) are properly contained in any larger elementary repeat.

3.2   Algorithm Description

Our algorithm for finding elementary repeats follows straight from our definition, and is easiest to describe assuming that $f = 2$. Let x be the first l-mer to be encountered twice, at positions i and j ($i < j$). x is potentially an elementary repeat (it is long enough and occurs at least 2 times in the genome), but it may not be maximal as required by condition (4).

Now let x_k be the l-mer starting k bases after the second occurrence of x (at position $j + k$). As we continue to scan l-mers we look for the smallest k such that one of three conditions occurs:

1. x_k is not equal to the l-mer at position $i + k$ (the kth l-mer after the first occurrence of x). In this case $x_0 \circ \cdots \circ x_{k-1}$ occurs at both i and j and forms a maximal identical substring pair, hence by Lemma 2 cannot be properly contained in any elementary repeat.
2. x_k is equal to some $x_{k'}$, $0 \le k' < k$. If this is the case, then $x_0 \circ \cdots \circ x_k$ cannot be an elementary repeat, nor can it be contained in one, as that would violate condition (2) of the definition. Thus $x_0 \circ \cdots \circ x_{k-1}$ cannot be properly contained in any elementary repeat.
3. x_k has occurred more times than x_0. In this case, $x_0 \circ \cdots \circ x_k$ cannot be an elementary repeat as x_k violates condition (3) of the definition. Thus $x_0 \circ \cdots \circ x_{k-1}$ cannot be properly contained in any elementary repeat.

At this point, we have a string $s = x_0 \circ \cdots \circ x_{k-1}$ that cannot be properly contained in an elementary repeat – but that *might* itself be an elementary repeat. By our choice of k we have ensured it meets criteria (1) and (2), and that criteria (3) has not yet been violated – but one or more of the x_i might be found, independent of s, later in the scan. So s is a *potential* elementary repeat (and would be an elementary repeat if x_{k-1} were the final l-mer of the genome).

As we continue this scan, we may identify more instances of the potential elementary repeat s, as well as establish other families. Further, we can retroactively decompose s as needed. For example: suppose $s = x_0 \circ \cdots \circ x_m$ has been identified as a potential elementary repeat, and we later find a string $s' = x_0 \circ \cdots \circ x_{k-1}$ where $k \le m$ is the halting k-value from above. This could only happen if the base following x_{k-1} in s is different than that following x_{k-1} in s', meaning x_k occurs k bases after the start of s but not after the start of s'. Hence s cannot be an elementary repeat, as x_k occurs more times that x_{k-1} in G, which is forbidden by condition (3) . But the strings $s_1 = x_1 \circ \cdots \circ x_{k-1}$ and $s_2 = x_k \circ \cdots \circ x_m$ are still each potential elementary repeats.

Theorem 1. *The algorithm outlined in Figure 1 will find all substrings conforming to the definition of an elementary repeat.*

The proof of this is a straightforward induction argument based on the post-condition that, when finished with the ith iteration of the loop, all elementary repeats *with respect to the partial genome* $G[0 : i + l]$ have been identified.

3.3   Pattern Hunter Augmentation

To this point we have required that all members of a basic elementary repeat family be exact matches. To allow for sequence variation between elements, we incorporate the idea of the spaced seed [14,16]. Given a binary string s of length l and two genomic strings q and t, also of length l, we say that q *hits* t with respect to s if $q_i = t_i$ for all i such that $s_i = 1$. For example, for $s = 10011$, $q = AACAA$ hits $t = AAAAA$ (with respect to s) because they match at all positions where s is 1; $q = AACAA$ does *not* hit $t = TACAA$. Given this, we can now relax our definition of a "match" between two repeats from a requirement of identity to the following:

Definition 3. *Let s be a seed of width l and q be a sequence of length $n \geq l$. We say that sequence t matches q with respect to s if: (1) $|t| = n$; (2) for each i, $0 \leq i < n$, there exists a value j, $0 \leq j < n - l$ such that $j \leq i < j + l$ and the substring $q[j : j + l]$ hits $t[j : j + l]$ with respect to seed s.*

In other words, two repeats are a match (hence in the same family) if every base in each string is covered by a substring that will hit a substring in the opposing sequence with respect to a seed. We can further expand this definition to allow for the use of multiple seeds, though have not yet explored this possibility.

Given this definition, it is a short step to relax the substring quality condition in our algorithm to that of a seed-based match. In relaxing the assumption of sequence identity we do toss away our algorithmic guarantee of finding all basic elements: as with BLAST HSP searches or general sequence alignment, badly placed mutations will defeat the seed and result in false negatives. But use of these seeds does decrease the probability of such occurrences, and PatternHunter has demonstrated that the appropriate seed combination can achieve very high sensitivity.

4   Results

The purpose of RAIDER is to provide a fast method for identifying elementary repeats in a manner robust to sequence variation. We were unable to obtain alternative tools for solving that problem [8,9,23], so comparisons have been made against RepeatScout – judged to be an effective tool for *de novo* repeat identification on assembled sequences [20]. This is not an ideal comparison, as the two tools are solving somewhat different problems (identification of elementary repeats v. repeat consensus library construction). It is, however, a reasonable comparison for the purpose of demonstrating the potential of RAIDER and the effect of the spaced seed structure. The RECON [3] and PILER [5] tools were also investigated, but did not provide any better results than RepeatScout and are not reported here.

```
SpliceFamily(Family F, l-mer x)
# Input: A family F and an l-mer x belonging to F
# Output / effect: Modified family F, creates new family F'
    F' ← family created from x and all subsequent l-mers in F
    F ← truncation of F to the l-mers preceding x
    return F'


ExactERSearch(sequence G, integer l, integer f):
# Input: Genomic sequence G, minimum length l, minimum family size f
# Output: A set of elementary repeat families
    F ← NULL, c ← 0       # F: Family currently under investigation
                          # c: An index into the family sequence
    for i ← 0 to |S| - l:  # Scan the l-mers of G in sequential order
        p ← S[i : i + l]       # Get the i-th l-mer of G (c-th l-mer of F)
        if F != NULL:
            r ← the c-th l-mer of F
            if p does not match r:    # Family is not a elementary repeat
                SpliceFamily(F, r)
                F ← NULL
            else: c = c + 1
        if count(p) = 2:              # Second instance of p -- new family
            q ← G[i - 1 : i + l - 1]
            if count(q)=2 and previous(q)=previous(p)-1:
                assign p to family(q)
            else: create new family, starting with l-mer p
        else if count(p) > 2:             # p must belong to a family
            if p is first l-mer in family(p):
                F ← family(p)
                c ← 1
            else if F = NULL or family(p) != F:
                F ← SpliceFamily(family(p), p)
                c ← 1
        if F != NULL and p is the last l-mer in F:
            F ← NULL
    return all families F containing at least f distinct elements
```

Fig. 1. Outline of the algorithm for elementary repeat extraction. Note we assume three functions for l-mer p and position i: $count(p)$ returns the number of occurrences of p in the first i bases of the G; $previous(p)$ returns the last occurrence of p before coordinate i, and $family(p)$ returns the family containing p. All tree functions can be implemented through standard data structures supporting constant time inspection and modification (given a fixed value of l), though details regarding the maintenance of those structures are omitted here.

Table 1. Runtime and memory usage on selected genomic sequences. Chromosomes marked with a * were modified to eliminate through base shuffling all repeats other than members of selected families. A ** indicates RAIDER was run with a pre-filter removing all low-frequency l-mers from consideration – reducing memory usage at a linear runtime penalty. We use an empirical measure of runtime based on the Linux time utility to precisely measure user time elapsed during execution of the process. No runtime is presented for RepeatScout on the full human genome as we were unable to run the tool to completion on the input (terminating on a "Unable to allocate - 1137348258 bytes for the sequence"). All tests were run on Redhat Linux using a 2.4 GHz Intel Xeon E5620 CPU and 128 GB RAM. RAIDER is coded in C++ using the C++11 standard and compiled with the GNU gcc 4.7 compiler.

Target	Tool	Runtime (s)	Memory (Gb)
Human chr. 22*	RAIDER	113	2.17
49691432 bp	RepeatScout	759	0.29
Mouse chr. 19*	RAIDER	185	3.39
61431566 bp	RepeatScout	3877	1.51
C. Elegans chr. I*	RAIDER	48	1.02
15072423 bp	RepeatScout	494	0.45
C. Elegans chr. X*	RAIDER	55	1.18
17718866 bp	RepeatScout	803	0.22
Human chr. 1	RAIDER	710	15.64
247299719 bp	RAIDER**	1791	3.01
	RepeatScout	5440	4.24
Human genome	RAIDER**	22830 ($\approx$ 6.3 hours)	29.49

Benchmark Data: Tests were conducted on the human genome as a whole, human chr. 1 on its own, and *modified* versions of human chr. 22, mouse chr. 19, and *c. elegans* chr. I and X. Specifically: for each of these last four chromosomes, we selected a series of known repeat locations and shuffled the genomic sequences between them. Thus we preserve repeat structure and sequence size but remove other repeats – eliminating unannotated repeats that might be discovered by our new approach and incorrectly labeled false positives.

Resource Usage. Table 1 contains the runtime and memory usage for each of the targets, showing a considerable improvement in runtime of RAIDER over RepeatScout, but at a cost in memory requirements. Due to this cost, large inputs require a pre-scan that filters out low-frequency l-mers (which are subsequently ignored in the algorithm described in Figure 1). We can see on human chromosome 1 the difference made by the pre-scan (increasing runtime from 719s to 1791s, still an 67% improvement over RepeatScout, but reducing memory requirements by over 80% (28%) as compared to standard RAIDER (RepeatScout)). The pre-scan was necessary to complete the full Human Genome on our server (with 128 Gb memory), but allows for completion of that entire sequence in under 6.4 hours. We were unable to complete a run of RepeatScout

Table 2. Results of applying RAIDER, with four different spaced seeds (denoted S_1, S_2, S_3, and S_4), and RepeatScout to the four sequences. Consensus coverage reports the % of bases in the RepBase [10] consensus sequences that are covered by the tool output in a BLAST alignment, reflecting the sensitivity of the tool. RepeatScout output is run through RepeatMasker before the BLAST comparison, though reported runtime does *not* include RepeatMakser runtime. Seed patterns are: $S_1 = 1^{24}$ (indicating 24 consecutive 1s – a basic 24-mer match strategy), $S_2 = 1^5 0 1^7 0 0 1^7 0 1^5$, $S_3 = 1^2 0^3 1^3 0^4 1^4 0^5 1^5 0^6 1^6 0^7 1^7$, and $S_4 = 1^4 0^8 1^5 0^{17} 1^6 0^{32} 1^7 0^{64} 1^8$.

Sequence	Method	Time (s)	consensus coverage (%)
human char 22	RAIDER: S_1	75	0.326
	RAIDER: S_2	77	0.315
	RAIDER: S_3	116	0.689
	RAIDER: S_4	192	0.84
	RepeatScout	2344	0.777
mouse chr 19	RAIDER: S_1	113	0.131
	RAIDER: S_2	130	0.218
	RAIDER: S_3	169	0.21
	RAIDER: S_4	303	0.307
	RepeatScout	3877	0.539
c. elegans chr I	RAIDER: S_1	27	0.622
	RAIDER: S_2	30	0.811
	RAIDER: S_3	43	0.582
	RAIDER: S_4	97	0.794
	RepeatScout	1329	0.944
c. elegans chr X	RAIDER: S_1	33	0.485
	RAIDER: S_2	39	0.393
	RAIDER: S_3	54	0.587
	RAIDER: S_4	89	0.558
	RepeatScout	1184	0.604

on the full human genome (encountering a memory allocation error resulting in a crash of the tool).

Result Quality. In Table 2 we look at result quality using RAIDER with different seeds, as compared to RepeatScout. This is a questionable comparison, as RAIDER is finding only basic elements while RepeatScout is searching for full repeats. Further, RepeatScout has the advantage of being paired with Repeat-Masker to produce its output (thus benefiting from a full library-based search), while RAIDER is being used stand-alone. (Note that reported runtime for RepeatScout does *not* include RepeatMasker runtime.) We quantify the results by BLASTing the output against the RepBase repeat consensus sequence file [10] for the query sequence, looking at the coverage of those consensus sequences as an indication of the relative sensitivity. (The specificity was near-perfect in all cases, and not reported here.) Seeds used with RAIDER were chosen arbitrarily;

a more formal investigation is required to find the best seed patterns, and we are currently in the process of generalizing the tool to allow for multiple seeds.

5 Discussion

We have implemented a novel linear time algorithm for finding elementary repeats in a genomic sequence that can be augmented using the PatternHunter spaced seed strategy. Testing of this tool proves that RAIDER is considerably faster than RepeatScout, and is able to come within reasonable distance of known repeat consensus coverage (as compared to the RepeatScout + RepeatMasker combination) in a fraction of the time despite being at a sizable disadvantage with respect to that metric. We have also demonstrated the validity of the method for incorporating the spaced seed strategy, and that variation in seed structure does make a significant difference to result quality (Table 2).

RAIDER is a work in progress, and there are some obvious holes in the strategy we are in the process of addressing. First: while there is some use in finding these basic elements, we would like to expand them into full consensus sequences usable as RepeatMasker library. Doing so appears to be a scaled down version of the sequence assembly problem, and we are currently testing software to address it. Second: in this initial version of RAIDER we allow for only a single spaced seed, and choose it arbitrarily. We are currently looking into the expansion of the method to accommodate multiple seeds; this strategy has proved effective in the original PatternHunter application [14,16]. Finally: we are looking at improving memory usage. While memory requirements are not so large as to prevent the application of RAIDER to the human genome, clearly some improvement in that area would be useful. Use of the Google SparseHash [6] introduce an unacceptable increase in runtime, but experimentation with our own implementation has shown some promise.

Acknowledgments. The research was conducting under funding from the National Science Foundation, Grant 0953215.

References

1. Smit, A.F.A., Hubley, R., Green, P.: RepeatMasker Open-1.0 (1996-2010), http://www.repeatmasker.org
2. Altschul, S., Madden, T., Schaffer, A., Zhang, J., Zhang, Z., Miller, W., Lipman, D.: Gapped BLAST and PSI-BLAST: a new generation of protein database search programs. Nucleic Acids Research 25(17), 3389 (1997)
3. Bao, Z., Eddy, S.R.: Automated de novo identification of repeat sequence families in sequenced genomes. Genome Research 12(8), 1269–1276 (2002)
4. Bergman, C.M., Quesneville, H.: Discovering and detecting transposable elements in genome sequences. Briefings in Bioinformatics 8(6), 382–392 (2007)
5. Edgar, R.C., Myers, E.W.: PILER: identification and classification of genomic repeats. Bioinformatics 21(suppl. 1), i152–i158 (2005)

6. Google: sparsehash - An extremely memory-efficient hash_map implementation - Google Project Hosting, http://code.google.com/p/sparsehash/
7. Hardison, R.C.: Covariation in Frequencies of Substitution, Deletion, Transposition, and Recombination During Eutherian Evolution. Genome Research 13(1), 13–26 (2003)
8. He, D.: Using suffix tree to discover complex repetitive patterns in DNA sequences. In: Conference Proceedings: ... of Annual International Conference of the IEEE Engineering in Medicine and Biology Society, vol. 1, pp. 3474–3477. IEEE Engineering in Medicine and Biology Society (2006)
9. Huo, H., Wang, X., Stojkovic, V.: An Adaptive Suffix Tree Based Algorithm for Repeats Recognition in a DNA Sequence. Bioinformatics and Bioengenierring, 181–184 (2009)
10. Jurka, J., Kapitonov, V.V., Pavlicek, A., Klonowski, P., Kohany, O., Walichiewicz, J.: Repbase Update, a database of eukaryotic repetitive elements. Cytogenetic and Genome Research 110(1-4), 462–467 (2005)
11. Karro, J.E., Peifer, M., Hardison, R.C., Kollmann, M., von Grünberg, H.H.: Exponential decay of GC content detected by strand-symmetric substitution rates influences the evolution of isochore structure. Molecular Biology and Evolution 25(2), 362–374 (2008)
12. Kurtz, S., Choudhuri, J.V., Ohlebusch, E., Schleiermacher, C., Stoye, J., Giegerich, R.: REPuter: the manifold applications of repeat analysis on a genomic scale. Nucleic Acids Research 29(22), 4633–4642 (2001)
13. Lander, E.S., et al.: Initial sequencing and analysis of the human genome. Nature 409(6822), 860–921 (2001)
14. Li, M., Ma, B., Kisman, D., Tromp, J.: Patternhunter II: highly sensitive and fast homology search. Journal of Bioinformatics and Computational Biology 2(3), 417–439 (2004)
15. Li, R., Ye, J., Li, S., Wang, J., Han, Y., Ye, C., Wang, J., Yang, H., Yu, J., Wong, G.K.S., Wang, J.: ReAS: Recovery of ancestral sequences for transposable elements from the unassembled reads of a whole genome shotgun. PLoS Computational Biology 1(4), e43 (2005)
16. Ma, B., Tromp, J., Li, M.: PatternHunter: faster and more sensitive homology search. Bioinformatics (2002)
17. Pevzner, P.A., Tang, H., Tesler, G.: De novo repeat classification and fragment assembly. Genome Research 14(9), 1786–1796 (2004)
18. Price, A.L., Jones, N.C., Pevzner, P.A.: De novo identification of repeat families in large genomes. Bioinformatics 21(suppl. 1), i351–8 (2005)
19. Saha, S., Bridges, S., Magbanua, Z.V., Peterson, D.G.: Computational Approaches and Tools Used in Identification of Dispersed Repetitive DNA Sequences. Tropical Plant Biology 1(1), 85–96 (2008)
20. Saha, S., Bridges, S., Magbanua, Z.V., Peterson, D.G.: Empirical comparison of ab initio repeat finding programs. Nucleic Acids Research 36(7), 2284–2294 (2008)
21. Zabala, G., Vodkin, L.: Novel exon combinations generated by alternative splicing of gene fragments mobilized by a CACTA transposon in Glycine max. BMC Plant Biology 7, 38 (2007)
22. Zerbino, D.R., Birney, E.: Velvet: algorithms for de novo short read assembly using de Bruijn graphs. Genome Research 18(5), 821–829 (2008)
23. Zheng, J., Lonardi, S.: Discovery of repetitive patterns in DNA with accurate boundaries ... (2005)
24. Zhi, D., Raphael, B.J., Price, A.L., Tang, H., Pevzner, P.A.: Identifying repeat domains in large genomes. Genome Biology 7(1), R7 (2006)

A Probabilistic Model Checking Analysis of the Potassium Reactions with the Palytoxin and Na$^+$/K$^+$-ATPase Complex

Fernando Braz[1], João Amaral[1], Bruno Ferreira[1],
Jader Cruz[2], Alessandra Faria-Campos[1], and Sérgio Campos[1]

[1] Department of Computer Science
[2] Biochemistry and Immunology Department
Universidade Federal de Minas Gerais
Av. Antônio Carlos, 6627, Pampulha, 30123-970, Belo Horizonte, Brazil
{fbraz,joaosale,bruno.ferreira,alessa,scampos}@dcc.ufmg.br,
jcruz@icb.ufmg.br

Abstract. In this paper, Probabilistic Model Checking (PMC) is used to model and analyze the effects of the palytoxin toxin (PTX) in cell transport systems, structures responsible for exchanging ions through the plasma membrane. The correct behavior of these systems is necessary for all animal cells, otherwise the individual could present pathologies. We have developed a model which focuses on potassium and cell energy related reactions, due to the known inhibitory effect of potassium on PTX action and the ATP role in its transportation. We have used PMC to estimate state probabilities and use the Goldman-Hodgkin-Katz equation to measure the induced current created by ion exchange. Our model suggests that as the concentration of external potassium increases, ion exchange occurs against its electrochemical gradient, despite the PTX effect. This suggests that potassium could be used to inhibit PTX action. PMC allowed us to further characterize the system dynamics.

Keywords: Probabilistic Model Checking, Systems Biology, Sodium-Potassium Pump, Palytoxin, Ion Channels Blockers and Openers.

1 Introduction

Probabilistic Model Checking (PMC) is a computational automated procedure to model and analyze complex systems that present non-deterministic and dynamic behavior. These stochastic characteristics are difficult to handle however frequently appear once we model real systems. The system description is modelled as a stochastic process such as Markov chains [14,20].

This procedure exhaustively and automatically explores the state space of a model, verifying if it satisfies properties given in probabilistic temporal logics, such as Continuous Stochastic Logic (CSL). Properties can be expressed as, for example, "the probability that a particular reaction occurs is at least 10%". Properties can offer valuable insight over model behavior [8,17].

J.C. Setubal and N.F. Almeida (Eds.): BSB 2013, LNBI 8213, pp. 181–193, 2013.
© Springer International Publishing Switzerland 2013

PMC can be directly applied to study biological systems which show probabilistic behavior, common at the cellular level. Chemical reactions and biological processes might occur, depending on the concentration of ligands (ions and molecules), and environmental and cellular conditions. PMC can be used to improve our understanding of these systems, complementary to others methods, such as stochastic and deterministic simulations, which present local minima problems that PMC avoids due to its exhaustive approach [16,15].

In this work, we present and evaluate a stochastic PMC model of the sodium-potassium pump (or Na^+/K^+-ATPase), an active cell transport system that exists in animal cells. The pump is important to several biological processes, such as cell volume control and heart muscle contraction. Its irregular behavior can be related to several diseases and syndromes, such as hypertension and Parkinson's disease, and it is one of the main targets of toxins and drugs [3].

In previous works, the pump has been exposed to a deadly toxin called palytoxin (PTX), which binds to the pump and disrupts its regular behavior. This has been done in order to understand the effects of PTX interactions with the pump. Our current model describes potassium (K) and cell energy related reactions since potassium has a known inhibitory effect on PTX and cell energy plays a major role on ion exchange [24].

We have used PMC to calculate state probabilities, which has allowed us to used the Goldman-Hodgkin-Katz (GHK) flux equation to measure the induced current created by ion exchange. Our model suggests that as the concentration of $[K^+]^o$ increases, the direction of ion exchange is reversed.

This suggests that $[K^+]^o$ could inhibit PTX action, which is a known inhibitory property of potassium on the PTX-pump complex [24]. The role of potassium could be further investigated in order to research novel methods to inhibit PTX action. PMC allowed us to further study the PTX-pump dynamics.

2   Background

This section describes the basic background on transmembrane ionic transport systems, namely ionic pumps and ion channels. Several aspects are discussed, such as their cycle, and associated diseases and syndromes.

2.1   Transmembrane Ionic Transport Systems

Animal cells contain structures called transmembrane ionic transport systems, which are responsible for ion exchange between the sides of the cell. The difference in charges and concentrations between ions creates an electrochemical gradient, which is essential for cells to perform their functions properly. Ionic transport systems are responsible for the maintenance of this gradient [2].

There are two types of transport systems: ion channels — a passive transport system which does not consume energy to promote ion exchange and ionic pumps — an active transport system that uses energy in the form of Adenosine Triphosphate (ATP) to perform ion exchange. Ion channels depend on the concentration gradient of the ions to be transported, moving them down their electrochemical gradient. Ionic pumps exchange ions against a concentration gradient [18], using ATP energy to do so. Once open, ion channels rapidly diffuse ions, allowing abrupt changes in ion concentration. Ionic pumps, on the other hand, exchange ions very slowly, allowing subtle changes in ion concentration.

Cell transport systems, such as Na$^+$/K$^+$-ATPase, are involved in several biological processes such as cellular volume control, nerve impulse and coordination of heart muscle contraction. These systems are one of the main targets in research for discovery and development of drugs, since its irregular behavior is associated with several diseases, such as hypertension, seizures, cystic fibrosis and Parkinson's disease.

Ion channels and ionic pumps allow only the passage of specific ions such as sodium (Na^+), potassium (K^+) and calcium (Ca^2+). For ionic pumps, the passage of ions can be viewed as two gates, one internal and one external, which open or close based on different factors, like chemical signals [2].

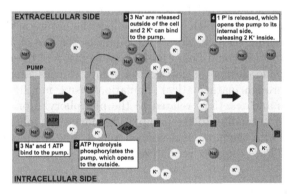

Fig. 1. The Sodium-Potassium Pump. Adapted from [25]

One example is the sodium-potassium pump or Na$^+$/K$^+$-ATPase (Figure 1). This pump is responsible for exchanging three sodium ions from the intracellular medium (rich in potassium and poor in sodium) for two potassium ions from the extracellular medium (poor in potassium and rich in sodium).

This pump can be in two major states: open to the inside of the cell, or open to the outside. The pump cycle starts with three sodium ions binding to the pump when its open to the intracellular side (first step of Figure 1). An ATP binds to the pump, which is followed by it is hydrolysis (second step). This breaks the ATP into two molecules, one of phosphate (P$_i$), which remains bound to the pump, and another of Adenosine Diphosphate (ADP), which is released inside the cell. This also causes the pump to release the sodium ions outside (third step). Two potassium ions in the outside bind to the pump, which are released in the intracellular side, as well as the phosphate (fourth and final step) [2].

Due to their role in the nervous system, ion transport systems are affected by neurotoxins [2]. One of the toxins that affects these structures is the palytoxin (PTX, or [PTX]^o for extracellular PTX concentration), a deadly toxin found in corals of the *Palythoa toxica* species. PTX disturbs the Na^+/K^+-ATPase, modifying its behavior to the one of an ion channel, which means that the pump transfers ions down their electrochemical gradient, instead of against it [3].

Ion channels and ionic pumps usually are investigated using experimental results in laboratory benches, which are expensive for both financial and time resources. In order to minimize these costs, different types of mathematical and computational methods have been employed, including sets of ordinary differential equations (ODEs) and Gillespie's algorithm for stochastic simulations [11]. Despite their ability to obtain valuable information, simulations do not cover every possibly situation, and might never search certain regions of the state space, therefore possible overlooking some events, such as ion depletion.

3   Related Work

3.1   Experimental and Simulational Techniques

The authors of [3] investigated PTX and its interactions with the Na^+/K^+-ATPase. They have discovered that PTX modifies the nature of the pump after binding to it, changing the behavior of the pump to the one of an ion channel. They suggest that PTX could be an useful tool to discover the control mechanisms for opening and closing the gates of ion pumps. This is later visited by the authors of [23] through mathematical simulations using non-linear ODEs and considering states and reactions related to the phosphorylation process (phosphate binding and unbinding to the pump). The potassium inhibitory effect on PTX interactions with the pump is described in [24]. The complete model of the PTX-Na^+/K^+-ATPase complex is analyzed in [22].

3.2   Model Checking

The authors illustrate in [16] the application of PMC to model and analyze different complex biological systems, for example the signaling pathway of Fibroblast Growth Factor (FGF), a family of growth factors involved in healing and embryonic development. The analysis of other signaling pathways such as MAPK and Delta/Notch can be seen in [15].

The use of PMC is demonstrated also in [13], where the authors examine and obtain a better understanding of mitogen-activated kinase cascades (MAPK cascades) dynamics, biological systems that respond to several extracellular stimuli, e.g. osmotic stress and heat shock, and regulate many cellular activities, such as mitosis and genetic expression.

The main tools used in the formal verification of biological systems that are related to this work are PRISM [17], BioLab [9], Ymer [27] and Bio-PEPA [7].

We have used PRISM for several reasons, which include: exact PMC in order to obtain accurate results; Continuous-time Markov Chain (CTMC) models, suited for our field of study; rich modeling language that allowed us to build our model and finally property specification using Continuous Stochastic Logic (CSL), which is able to express qualitative and quantitative properties.

4   The Model

4.1   Na$^+$/K$^+$-ATPase

The model is written in the PRISM model checker language [17]. It consists of PRISM modules for each of the ligands (K and ATP), one main module for the pump, and one auxiliary module which defines the speed of each reaction.

Each ligand module contains a variable to store the number of molecules, e.g. atpIn for ATP. Modules are composed of PRISM commands (or transitions), which are responsible for updating the model. They represent reactions and are responsible for changing the number of molecules or the state of the pump.

A PRISM command uses the following structure: [sync] conditions → rate : update, where the conditions must be observed for the update to occur at a given rate. The sync is used to synchronize multiple commands and it is useful, for example, to define the speed of the reactions.

```
module k
  kIn  : [0..(KI+KO)] init KI; // Number of K inside cell
  kOut : [0..(KI+KO)] init KO; // Number of K outside cell
  // reaction 2: 2kIn + E1 <-> _K2_Eocc
  [r2]  kIn >= kFlow2        -> pow(kIn,2) : (kIn' = kIn - kFlow2);
  [rr2] kIn <= (KI+KO-kFlow2) -> 1         : (kIn' = kIn + kFlow2);
endmodule
module pump
  E1      : [0..1] init 1; // pump open to its internal side
  _K2_Eocc : [0..1] init 0; // pump occluding two potassium ions
  // reaction 2: 2kIn + E1 <-> _K2_Eocc
  [r2]  ki != 0 & E1 = 1 & _K2_Eocc = 0 -> 1 : (E1' = 0) & (_K2_Eocc' = 1);
  [rr2] ki != 0 & E1 = 0 & _K2_Eocc = 1 -> 1 : (E1' = 1) & (_K2_Eocc' = 0);
endmodule
// base rates
const double r2rate  = 1.00 * pow(10,2);
const double rr2rate = 1.00 * pow(10.0,-1);
module base_rates
  [r2]  true -> r1rate  : true;
  [rr2] true -> rr1rate : true;
endmodule
```

Fig. 2. Na$^+$/K$^+$-ATPase PRISM Model

This model does not include PTX because its interactions with the pump are an extension presented in the next subsection. A fragment of the model is shown in Figure 2 and its complete version can be seen online [1].

The conditions for a module command to be executed usually are lower and upper bounds, i.e. there must be at least one molecule for a binding reaction, or the pump must be in a particular state. The list of reactions can be found in [24] and in the comments of our model [1].

We have used the construct pow(x,y) for power functions from PRISM to represent the law of mass action, explained further below (Discrete Chemistry). For example, a reaction involving two extracellular potassium ions would have a transition rate pow(kIn,2).

The main module controls the pump, keeping track of its current sub-state. The sub-states are boolean vectors, where only one position can and must be true. There are also several global variables which are used across the whole model, such as ligand concentrations and pump volume.

The Albers-Post kinetic model [21] represents the pump cycle (Figure 3). The pump can be in different sub-states, which change depending on reactions involving K and ATP. Figure 3 shows the PTX extension model (discussed later).

The pump can be open, allowing ion exchange, or closed, blocking ion movement and possibly occluding ions. An ATP molecule can bind to the pump in its high or low affinity binding sites. The pump can contain two potassium ions. The reactions are bidirectional and their rates were obtained in [24].

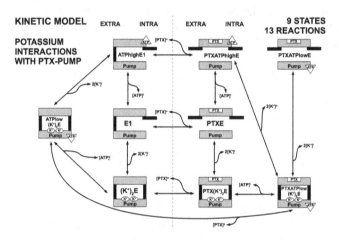

Fig. 3. Kinetic Model. The kinetic model for the coupling and uncoupling of PTX to the pump describes all the sub states (9) and reactions (11). The left side is the classical Albers-Post model [21], which describes the regular behavior of the pump, while the right side describes the PTX related states and reactions [24].

Previously, a PMC model of the pump was described in [10]. PTX was included in the model in [4], where disturbances caused by the toxin in cell energy related reactions were studied. A model which focused on sodium and potassium related reactions was described in [5]. It revealed that sodium enhances PTX action, while potassium probably inhibits it. Since the toxin is found in marine species, the sodium inhibitory effect is not a coincidence.

Palytoxin Extension. The palytoxin model is an extension of the Na^+/K^+-ATPase model, described in the previous section (Figure 3). It is based on the description of [24] and [3]. This extension consists of: one additional molecule module (PTX) which controls its flow; additional reactions in each of the already present modules; and additional sub-states and transitions for the pump module. Initial concentrations for $[PTX]^o$ and stochastic rates for reactions were obtained in [24]. The six additional sub-states correspond to the pump bound to PTX, when the pump is open to both sides behaving like an ion channel.

4.2 Discrete Chemistry and the Law of Mass Action

Our model is composed of potassium ions, ATP molecules and the pump itself, which can interact with each other through several chemical reactions. There is one additional molecule (PTX) in the palytoxin extension of this model.

The ligand concentrations are discrete variables, instead of continuous functions. Therefore, we have converted the amount of initial concentration of molecules from molarity (M) to number of molecules. The stochastic rates for forward and backward transitions and the ligands concentrations ($[ATP]^i = 0.005$ M, $[K]^i = 0.127$ M and $[K]^o = 0.010$ M) have been obtained in [24,6].

In order to convert the initial amount of molecules given in molarity ($[X]$) into quantities of molecules ($\#X$), we have used the following biological definition [2]:

$$\#X = [X] \times V \times N_A \tag{1}$$

where V is the cell volume and N_A is the Avogadro constant (6.022×10^{23} mol^{-1}).

The law of mass action states that a reaction rate is proportional to the concentration of its reagents. Therefore, we take into account the ligands concentrations in our model. Considering the discrete chemistry conversion discussed and the palytoxin binding to the pump:

$$E_1 + \text{PTX} \overset{rp_1'}{\longrightarrow} \text{PTX} \sim E \tag{2}$$

the final rate rp_1 is given as follows:

$$rp_1 = rp_1' \times \#(E_1) \times \#(\text{PTX}) \tag{3}$$

5 Results

This section begins describing the model parameters and complexity, followed by a discussion on the scenarios that have been studied. The state and transition probabilities are discussed in Section 5.2, and the induced current created by ion exchange is covered in Section 5.3.

5.1 Parameters and Model Complexity

We can explore our model by changing its four dimensions: $[PTX]^o$ (extracellular PTX concentration), $[K^+]^o$ (extracellular potassium concentration), $[ATP]^i$ (intracellular ATP concentration) and pump volume. Each dimension represents one aspect of the model, and can be changed to modify its behavior.

These parameters influence directly the complexity of the model (number of states, transitions and topology), and the time to build and verify model properties, as it can be seen in Table 1. The machine used to perform experiments is an Intel(R) Xeon(R) CPU X3323, 2.50GHz and has 17 GB of RAM memory.

The analysis is restricted to only one pump. As a consequence, it would not be realistic to model a large volume because in the real cell it is shared between several pumps and other cellular structures, not limited to pumps.

Our analysis is focused on single channels, therefore our abstraction reduces the cell volume to one pump and its surroundings. We achieve this by maintaining

Table 1. Model complexity as states and transitions translates into the size of the model state space. Properties include state and transition rewards (T_{State} and T_{Rate}).

Scenario	States	Transitions	T_{State}	T_{Rate}
Control	220	620	119.165 s	129.201 s
High $[K^+]^o$	568	1640	914.083 s	913.998 s
High $[ATP]^i$	568	1640	305.436 s	303.244 s

the proportions between all interacting components. Our dimension for cellular volume is called pump volume. Even though those values are many orders of magnitude smaller than the real values, they still represent proper cell behavior.

However, we have created three scenarios, which are compared with each other for different analysis. The Control scenario presents the regular physiological conditions of the pump. In the High Potassium ($[K^+]^o = 0.100$ M) and High Adenosine Triphosphate ($[ATP^+]^i = 0.050$ M) scenarios, the concentrations of extracellular potassium and intracellular ATP are both increased ten times.

Diseases can change the ATP concentration. Literature has reported cases on the matter (although we were not able to find a quantitative study), for example, a case of Huntington's disease [19]. This applies to potassium, which can reach increased concentrations in diseases such as hyperkalemia [26].

5.2 PTX-pump State Change for Different Scenarios

In order to observe the probability of PTX and non-PTX related states over time, all states and rates were quantified using rewards. Figure 4 shows the reward of the sub-state `PTXATPhighE`, where the pump is open to both sides, bound to PTX and an ATP in its high affinity site. Rewards are accumulated as time is spent in corresponding states as this is a continuous time model [14].

State Reward PRISM Model	*Accumulated State Reward Property*

`rewards "ptxatphighe"`
`  (PTXATPhighE=1) : 1;`
`endrewards`
	R{"ptxatphighe"}=? [**C**<=T]
	What is the accumulated reward for the state `ptxatphighe` at time T?

Fig. 4. State Reward and Accumulated State Reward Property

Since the model now has rewards for each state, we are able to count the expected quantity of the accumulated reward associated with each sub-states over time. Using the operator **R** we are able to quantify the reward for some given event, for example the number of times the model was in sub-state `PTXATPhighE` and `PTXATPlowE`. The operator **C** allows to quantify accumulated rewards for a given time T, therefore we are able to observe rewards over time.

Considering the Control scenario for a single pump at instant T=100, the expected rewards associated with the sub-states `PTXATPhighE` and `PTXATPlowE` is 95.8456 and 4.1502, respectively. In other words, in 100 seconds, the pump is expected to be bound to PTX and ATP in its high and low affinity binding site 95.8485% and 4.1503% of the time.

Using other scenarios such as High $[K^+]^o$ and High $[ATP]^i$, we have found that for the cellular volume of 10^{-22} L there are sets of values for sub-state rewards. One set is associated with the Control and High $[ATP]^i$ scenarios, while the other with the High $[K^+]^o$ scenario. For example, in the High $[K^+]^o$ scenario, the expected rewards associated with the sub-states PTXATPhighE and PTXATPlowE change to respectively 0.0011 and 99.9964, or 0.00001% and 99.9978% of the time. These results have been summarized in Figure 5.

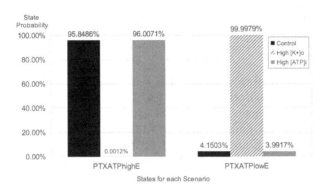

Fig. 5. Probability of States Responsible for PTX Induced Channels

This change in the most active state of the pump is important to understand the inhibitory effect of potassium on PTX. Each of these states have a different role in electric measurements of the induced current caused by ion exchange. Only after measuring the probability of states where the pump is open it was possible to perform these induced electric current measurements, since their probability is part of the equations, which are described below.

5.3   Induced Electric Current Measurements

The induced current carried by potassium ions across the membrane can be measured using the Goldman-Hodgkin-Katz (GHK) flux equation (divided into Equations 4 and 5), which describes the ionic flux as a function of transmembrane potential and potassium ion concentrations [12,2].

This equation allows studying quantitatively the behavior of the induced current under different conditions, such as the scenarios previously mentioned.

$$J_{ion} = P_{ion}\, z_{ion}^2\, \frac{F^2\, V_m}{R\, T}\, \frac{[ion]^i\, e^{\frac{z_{ion}\, F\, V_m}{R\, T}} - [ion]^o}{e^{\frac{z_{ion}\, F\, V_m}{R\, T}} - 1} \tag{4}$$

P_{ion} represents the permeability of the membrane for that ion and it is shown in Equation 5.

$$P_{ion} = \gamma_1[PTXE] + \gamma_2[PTXATPhighE] + \gamma_3[PTXATPlowE] \tag{5}$$

Descriptions, values and units for constants, such as V_m (transmembrane potential) and F (Faraday constant) are shown in Table 2. Further details on the GHK equation can be found in [24].

Table 2. The constants used in the Goldman-Hodgkin-Katz (GHK) flux equation

Constant	Description	Value	Unit
J_{ion}	Ionic induced current	—	Ampere$\times$ meter^{-2}
z_K	Valence of potassium ion	1.00	—
γ_1	PTXE proportionality	0.15	—
γ_2	PTXATPhighE proportionality	0.15	—
γ_3	PTXATPlowE proportionality	0.90	—
V_m	Transmembrane potential	-20.00	miliVolts
F	Faraday constant	96,485.00	Coulomb$\times$ mol^{-1}
R	Gas constant	8,314.00	Joules$\times$ Kelvin$^{-1}\times$ mol^{-1}
T	Temperature	310.00	Kelvin

Our PMC approach always obtains the real expected value for our model due to its exhaustive exploration of all states. This is particularly important for electric current measurements (Figure 6), because it depends on the probability of states PTXE (the PTX-pump complex), PTXATPhighE and PTXATPlowE (the PTX-pump complex with an ATP bound to its high and low affinity binding sites, respectively), as well as internal and external ion concentrations. These state probabilities and ion concentrations change for different scenarios.

Measurements are performed through instantaneous reward properties, which obtain the reward value precisely at T time. A negative induced current indicates normal potassium ion flux (two potassium ions go inside), while a positive one indicates an abnormal flux (two potassium ions go outside).

In the Control scenario, the induced current is positive ($52,547\frac{A}{m^2}$), which means that potassium ions are leaving the cell in favor of their electrochemical gradient due to the effect of PTX, which can disrupt regular cell behavior.

Electric Current Reward	*Instantaenous Current Reward Property*

```rewards "pos_jK"```
``` (jK>=0): jK;```
```endrewards```
	$\mathbf{R}\{\text{"pos\_jK"}\}=?\ [\ \mathbf{I}{=}\mathbf{T}\ ]$
	
	What is the expected instantaneous reward for the positive current pos_jK at time T?

Fig. 6. Electric Current Reward and Instantaenous Current Reward Property

In the High $[K^+]^o$ scenario, the current becomes negative ($-426,790\frac{A}{m^2}$) – potassium transport changes, having its direction reversed. Potassium ions are entering the cell against their electrochemical gradient, despite the effect of PTX.

The states PTXATPhighE and PTXATPlowE play a major role in the GHK flux equation. We have observed that their probability change from one scenario to the other. However, their proportionality coefficients are largely different ($\gamma_1 = 0.15$ for PTXATPhighE and $\gamma_2 = 0.9$ for PTXATPlowE). This change in the most active state and the increased concentration of [K$^+$]^o explain the sudden reversion of ion exchange, as both are components of the equation.

This suggests that high concentrations of [K$^+$]^o could inhibit PTX action, which is an already known inhibitory property of potassium on the PTX-pump complex [24]. The potassium role could be further studied in order to discover novel strategies to create agents to inhibit PTX action.

6   Conclusions and Future Work

In this work, the known inhibitory effect of potassium on PTX is further characterized. Our PMC approach has allowed us to explore the model, accurately calculating its probabilistic characteristics. We have measured the probability of every state and reaction of the pump, for three different scenarios.

These scenarios have been created to simulate different conditions for the pump, such as diseases or intoxication. The Control scenario presents the regular physiological conditions of the pump. In the High [K$^+$]^o and High [ATP]^i scenarios, as the names imply, the concentrations of extracellular potassium and intracellular ATP are increased ten times, respectively.

In the Control scenario, the most active state of the pump is the PTX-pump complex with an ATP molecule bound to its high affinity binding site (**PTXATPhighE**), approximately 95.84% of the time. In the High [K$^+$]^o scenario, the most active state shifts to a similar one, except that now ATP is bound to its low affinity binding site (**PTXATPlowE**), nearly 99.99% of the time.

We have used the Goldman-Hodgkin-Katiz (GHK) flux equation to measure the induced current caused by ion exchange. In the Control scenario, the induced current is positive, which means that potassium is moving in favor of its electrochemical gradient. In the High [K$^+$]^o scenario, the induced current becomes negative – potassium transport is reversed, despite the PTX action.

This suggests that [K$^+$]^o could inhibit PTX action, which is a known inhibitory property of potassium on the PTX-pump complex. Potassium role could be further investigated in order to research novel methods to inhibit PTX action. PMC allowed us to further characterize the PTX-Na$^+$/K$^+$-ATPase dynamics.

Future work include the validation of our results through wet lab experiments and the expansion of the model to the complete Albers-Post kinetic model. Furthermore, our model can be extended to other toxins or even drugs.

References

1. http://www.dcc.ufmg.br/~fbraz/bsb2013/
2. Aidley, D.J., Stanfield, P.R.: Ion channels: molecules in action. Cambridge University Press (1996)
3. Artigas, P., Gadsby, D.C.: Large diameter of palytoxin-induced Na/K pump channels and modulation of palytoxin interaction by Na/K pump ligands. J. Gen. Physiol. 123(4), 357–376 (2004)

4. Braz, F.A.F., Cruz, J.S., Faria-Campos, A.C., Campos, S.V.A.: A probabilistic model checking approach to investigate the palytoxin effects on the Na^+/K^+-ATPase. In: de Souto, M.C.P., Kann, M.G. (eds.) BSB 2012. LNCS, vol. 7409, pp. 84–96. Springer, Heidelberg (2012)
5. Braz, F.A.F., Cruz, J.S., Faria-Campos, A.C., Campos, S.V.A.: Palytoxin inhibits the sodium-potassium pump – an investigation of an electrophysiological model using probabilistic model checking. In: Gheyi, R., Naumann, D. (eds.) SBMF 2012. LNCS, vol. 7498, pp. 35–50. Springer, Heidelberg (2012)
6. Chapman, J.B., Johnson, E.A., Kootsey, J.M.: Electrical and biochemical properties of an enzyme model of the sodium pump. Membrane Biology (1983)
7. Ciocchetta, F., Hillston, J.: Bio-pepa: A framework for the modelling and analysis of biological systems. Theoretical Computer Science (2009)
8. Clarke, E.M., Emerson, E.A.: Design and synthesis of synchronization skeletons using branching-time temporal logic. In: Kozen, D. (ed.) Logic of Programs 1981. LNCS, vol. 131, pp. 52–71. Springer, Heidelberg (1982)
9. Clarke, E.M., Faeder, J.R., Langmead, C.J., Harris, L.A., Jha, S.K., Legay, A.: Statistical model checking in *BioLab*: Applications to the automated analysis of T-cell receptor signaling pathway. In: Heiner, M., Uhrmacher, A.M. (eds.) CMSB 2008. LNCS (LNBI), vol. 5307, pp. 231–250. Springer, Heidelberg (2008)
10. Crepalde, M., Faria-Campos, A., Campos, S.: Modeling and analysis of cell membrane systems with probabilistic model checking. BMC Genomics 12 (2011)
11. Gillespie, D.T.: Exact stochastic simulation of coupled chemical reactions. The Journal of Physical Chemistry 81(25), 2340–2361 (1977)
12. Hille, B.: Ion Channels of Excitable Membranes, 3rd edn. Sinauer Associates (2001)
13. Kwiatkowska, M., Heath, J.: Biological pathways as communicating computer systems. Journal of Cell Science 122(16), 2793–2800 (2009)
14. Kwiatkowska, M., Norman, G., Parker, D.: Stochastic model checking. In: Bernardo, M., Hillston, J. (eds.) SFM 2007. LNCS, vol. 4486, pp. 220–270. Springer, Heidelberg (2007)
15. Kwiatkowska, M., Norman, G., Parker, D.: Quantitative Verification Techniques for Biological Processes. In: Algorithmic Bioprocesses. Springer (2009)
16. Kwiatkowska, M., Norman, G., Parker, D.: Probabilistic Model Checking for Systems Biology. In: Symbolic Systems Biology, pp. 31–59. Jones and Bartlett (2010)
17. Kwiatkowska, M., Norman, G., Parker, D.: PRISM 4.0: Verification of probabilistic real-time systems. In: Gopalakrishnan, G., Qadeer, S. (eds.) CAV 2011. LNCS, vol. 6806, pp. 585–591. Springer, Heidelberg (2011)
18. Nelson, D.L., Cox, M.M.: Lehninger Principles of Biochemistry, 3rd edn. (2000)
19. Olah, J., Klivenyi, P., Gardian, G., Vecsei, L., Orosz, F., Kovacs, G.G., Westerhoff, H.V., Ovadi, J.: Increased glucose metabolism and ATP level in brain tissue of Huntington's disease transgenic mice. FEBS J. 275(19), 4740–4755 (2008)
20. Parker, D.: Implementation of Symbolic Model Checking for Probabilistic Systems. Ph.D. thesis, University of Birmingham (2002)
21. Post, R., Refyvary, C., Kume, S.: Activation by adenosine triphosphate in the phosphorylation kinetics of sodium and potassium ion transport adenosine triphosphatase. J. Biol. Chem. 247, 6530–6540 (1972)
22. Rodrigues, A.M., Infantosi, A.F., de Almeida, A.C.: Palytoxin and the sodium/potassium pump–phosphorylation and potassium interaction. Phys. Biol. 6(3), 036010 (2009)
23. Rodrigues, A.M., Almeida, A.C.G., Infantosi, A.F., Teixeira, H.Z., Duarte, M.A.: Model and simulation of Na+/K+ pump phosphorylation in the presence of palytoxin. Computational Biology and Chemistry 32(1), 5–16 (2008)

24. Rodrigues, A.M., Almeida, A.C.G., Infantosi, A.F., Teixeira, H.Z., Duarte, M.A.: Investigating the potassium interactions with the palytoxin induced channels in Na+/K+ pump. Computational Biology and Chemistry 33(1), 14–21 (2009)

25. Sadava, D., Heller, H., Orians, G., Purves, W., Hillis, D.: Life: The Science of Biology. Sinauer Associates (2006)

26. Sevastos, N., Theodossiades, G., Efstathiou, S., Papatheodoridis, G.V., Manesis, E., Archimandritis, A.J.: Pseudohyperkalemia in serum: the phenomenon and its clinical magnitude. Journal of Laboratory and Clinical Medicine 147(3), 139–144 (2006), http://www.sciencedirect.com/science/article/pii/S002221430500404X

27. Younes, H.L.S.: Ymer: A statistical model checker. In: Etessami, K., Rajamani, S.K. (eds.) CAV 2005. LNCS, vol. 3576, pp. 429–433. Springer, Heidelberg (2005)

False Discovery Rate for Homology Searches

Hyrum D. Carroll[1], Alex C. Williams[1],
Anthony G. Davis[1], and John L. Spouge[2]

[1] Middle Tennessee State University
Department of Computer Science
Murfreesboro, TN 37132, United States of America
Hyrum.Carroll@mtsu.edu, {acw4a,agd2q}@mtmail.mtsu.edu
[2] National Center for Biotechnology Information
Bethesda, MD 20894, United States of America
spouge@ncbi.nlm.nih.gov

Abstract. While many different aspects of retrieval algorithms (*e.g.*, BLAST) have been studied in depth, the method for determining the retrieval threshold has not enjoyed the same attention. Furthermore, with genetic databases growing rapidly, the challenges of multiple testing are escalating. In order to improve search sensitivity, we propose the use of the false discovery rate (FDR) as the method to control the number of irrelevant ("false positive") sequences. In this paper, we introduce BLAST$_{FDR}$, an extended version of BLAST that uses a FDR method for the threshold criterion. We evaluated five different multiple testing methods on a large training database and chose the best performing one, Benjamini-Hochberg, as the default for BLAST$_{FDR}$. BLAST$_{FDR}$ achieves 14.1% better retrieval performance than BLAST on a large (5,161 queries) test database and 26.8% better retrieval score for queries belonging to small superfamilies. Furthermore, BLAST$_{FDR}$ retrieved only 0.27 irrelevant sequences per query compared to 7.44 for BLAST.

1 Introduction

In response to a query, many database search algorithms (*e.g.*, BLAST [2]) return a sorted retrieval list of sequences with an E-value assigned to each sequence. Typically, each E-value is calculated from a statistical model of irrelevant database sequences and approximates the expected number of irrelevant sequences with a score equal to or better than the one calculated. Many algorithms truncate their retrieval lists at a uniform E-value threshold. We call this truncation procedure "uniform E-value thresholding". While many different aspects of BLAST have undergone rigorous examination, uniform E-value thresholding has not had the same scrutiny.

As computing potential and the sophistication of computer algorithms increase, so has the need to account for multiple testing. For homology searches, the query is compared against each sequence in the database independently, resulting in multiple tests. Performing multiple tests can give the perception of a

J.C. Setubal and N.F. Almeida (Eds.): BSB 2013, LNBI 8213, pp. 194–201, 2013.
© Springer International Publishing Switzerland 2013

more significant result than the data can support. False discovery rate (FDR) methods aim to control the proportion of irrelevant matches to address the issues that multiple testing introduces. They are widely used in microarray studies and virtually in all facets of genomic studies. Additionally, a FDR approach was recently used to aid in generating the DFam database [13].

Early efforts for managing the false positive rate aimed to control the Family-wise Error Rate (FWER), the likelihood of making one or more false discoveries. Due to the intrinsic nature of how the FWER is computed, FWER methods also provide control over the FDR. Four modern and traditionally-accepted FWER methods are the Bonferroni correction [4], the Holm step-down procedure [10], the Hochberg step-up procedure [9], the Hommel single-wise procedure [11]. The Bonferroni correction uses a uniform P-value threshold determined by a user-specified α (or P-value threshold) divided by the total number of performed tests. The Holm step-down procedure extends the Bonferroni correction by adding the rank of the ordered P-values to the total number of performed tests in the thresholding method. Like the Holm procedure, the Hochberg step-up process utilizes the rank in the thresholding method by looking for the P-value that is less than a user-specified α divided by the total number of performed tests in addition to the current P-value's rank. The Hommel single-wise procedure is similar in that it looks for the P-value for which all P-values with a higher rank are greater than a number proportional to α. Procedures designed to control only the FDR, such as the Benjamini-Hochberg procedure [3], are generally less conservative forms of measurement than FWER methods and never perform worse. The Benjamini-Hochberg method computes a threshold by multiplying the current P-value's rank by a user-specified α and dividing the result by the total number of performed tests.

In this paper, we explore the performance of $BLAST_{FDR}$, a BLAST variant that uses E-values to calculate the FDR. We demonstrate that $BLAST_{FDR}$ performs better than BLAST, in part by drastically decreasing the number of irrelevant sequences. The Methods section presents the implementation details of $BLAST_{FDR}$; the Results section details our testing procedures and their results. We conclude with a discussion of $BLAST_{FDR}$'s applicability.

The C++ source code for $BLAST_{FDR}$ and instructions are available at http://www.cs.mtsu.edu/~hcarroll/blast_fdr/.

2   Methods

BLAST accepts a sequence as a query to search for relevant matches in a specified database. Additionally, an E-value threshold may be supplied to BLAST. BLAST looks for all relevant matches between that query and the sequences in a database and then applies uniform E-value thresholding by ignoring all matches with an E-value above the specified value.

$BLAST_{FDR}$ extends version 2.2.27 of NCBI's BLAST algorithm by replacing uniform E-value thresholding with a one of the following algorithms: Bonferroni, Holm's step-down process, Hochberg's step-up process, Hommel's single-wise process, and Benjamini and Hochberg's method. The Bonferroni method calculates a threshold value for each sequence retrieved and considers the first k ranked sequences as significant that satisfy the following criterion: $P_k \leq \frac{\alpha}{m}$, where P_k is the P-value of the kth sequence and m is the size of the database searched. Because BLAST relies heavily on E-values instead of P-values, and given that E-value = P-value * m, we implemented the Bonferroni method as: $E_k \leq \alpha$ with E_k being the E-value of the kth sequence. Furthermore, the Holm method considers matches significant that meet the following criterion: $E_k \leq \frac{m\alpha}{m+1-k}$. Similarly, the Hochberg method takes a different approach by starting at the least likely match and working toward the best statistical score to consider the following matches as significant: $E_k \leq \frac{m\alpha}{m+1-k}$. The Hommel method also iterates from the least significant match to find the index k such that: $E_{m-k+j} > \frac{j\alpha}{k}$ for $j = 1, \ldots, k$, then uses k to consider the following matches significant: $E_k \leq \frac{m\alpha}{k}$. Finally, the Benjamini-Hochberg method iterates from the match with the best statistical score and uses the following criterion for significant matches: $E_k \leq k\alpha$.

Each match in BLAST is called a high scoring pair (HSP). A database sequence can have multiple HSPs. BLAST organizes all of the HSPs according to the database sequence to which they belong and maintains its internal data structures sorted by the best HSP per database sequence. This is problematic for applying the methods above. Consequently, BLAST$_{FDR}$ restructures the HSPs from sorted by sequence to sorted by individual scores before applying the threshold.

To determine retrieval efficacy, we leveraged the query sequences in the As-TRAL40 database [6]. Each sequence in the ASTRAL40 database has less than 40% sequence identity to the other sequences. More importantly, each sequence has been classified into a "superfamily". We only considered the queries that have at least one other superfamily member in the database. Matches with the sequences in the same superfamily are considered relevant matches. To avoid making erroneous assignments, we ignore matches that are not in the same superfamily as the query sequence. For irrelevant matches, we augmented this database 100-fold with random sequences drawn from the distribution of amino acids residues and length of sequences found in the original ASTRAL40 database.

We partitioned the augmented database into Training and Test databases. We sorted the queries by name, and assigned the 5,162 odd sequences to the Training database and the 5,161 even sequences to the Test database [1]. Additionally, we randomly selected 103 queries (2%) from the training dataset to use to evaluate which method to use. We refer to this subset as "Training-subset".

In this study, we utilize the Threshold Average Precision (TAP) [5] method as the evaluation criterion for retrieval efficacy. The TAP method calculates the median Average Precision-Recall with a moderate adjustment for irrelevant sequences just before the threshold. TAP values range from 0.0 for a retrieval with no relevant sequences to 1.0 for a search that retrieves all of the relevant sequences and only relevant sequences.

To determine the best performing method to use from the list above, we examined the retrieval performance for each one of them with $\alpha = \{0.0005, 0.005,$

Table 1. Average BLAST$_{FDR}$ TAP values using the Training-subset database

Method	α			
	0.0005	0.005	0.05	0.5
Bonferroni	0.163	0.170	0.198	0.199
Holm	0.163	0.170	0.198	0.199
Hochberg	0.081	0.088	0.102	0.150
Hommel	0.163	0.170	0.198	0.199
Benjamini-Hochberg	0.168	0.180	0.203	0.184

Table 2. Average BLAST$_{FDR}$ TAP values using the Training database

Method	α			
	0.0005	0.005	0.05	0.5
Benjamini-Hochberg	0.199	0.215	0.229	0.220

0.05, 0.5} using the Training-subset database. From these methods, we adopted the best performing one as the default threshold method in BLAST$_{FDR}$. We then evaluated that method with $\alpha = \{0.0005, 0.005, 0.05, 0.5\}$ using the entire training database. Finally, the best performing method with the best performing value of α was compared against BLAST using the Test database.

3   Results

To evaluate the performance of BLAST$_{FDR}$, we performed several experiments involving five different methods to account for multiple testing. We utilized an augmented version of the ASTRAL40 database (see the Methods section). We measure the performance in terms of the Threshold Average Precision (TAP) value.

First, we evaluated BLAST$_{FDR}$ with the following methods for determining the threshold for matches: Bonferroni correction, Holm step-down procedure, Hochberg step-up procedure, Hommel single-wise procedure and Benjamini-Hochberg. For each method, we set $\alpha = \{0.0005, 0.005, 0.05, 0.5\}$ on the Training-subset database (see Table 1). Of these methods, BLAST$_{FDR}$ with the Benjamini-Hochberg method received the best average TAP value of 0.203 and generally performed better than the other methods. Consequently, we adopted this method as the default for BLAST$_{FDR}$. For comparison purposes, BLAST received an average TAP value of 0.171 on the same database.

Table 3. Average TAP values for BLAST and BLAST$_{FDR}$

Database	BLAST	BLAST$_{FDR}$
Training-subset	0.171	0.203
Training	0.203	0.229
Test	0.198	0.226

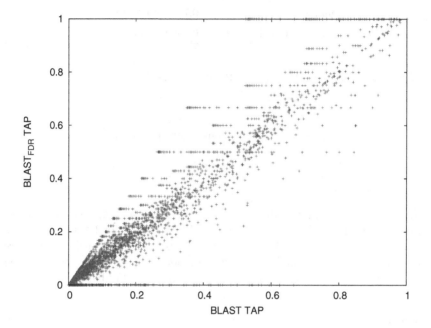

Fig. 1. TAP results for every query in the Test database

On the (full) Training database, we evaluated the same four α values for BLAST$_{FDR}$ using the Benjamini-Hochberg method (see Table 2). Of these parameters, BLAST$_{FDR}$ with $\alpha = 0.05$ received the best average TAP of 0.229 while BLAST received 0.203. Consequently, we adopted this α level as the default for BLAST$_{FDR}$.

We evaluated the efficacy of BLAST and BLAST$_{FDR}$ using the 5,161 query sequences in the Test database. Table 3 summarizes the results and Figure 1 details the TAP values for BLAST plotted against the TAP values for BLAST$_{FDR}$ for each of the queries. While BLAST received an average TAP value of 0.198, BLAST$_{FDR}$ earned an average TAP value of 0.226. In terms of irrelevant sequences, BLAST$_{FDR}$ retrieves an average of only 0.27 irrelevant sequences per query whereas BLAST retrieves 2,780% more with 7.44 per query. Finally, Figure 2 is a histogram of the E-values of sequences retrieved by BLAST that

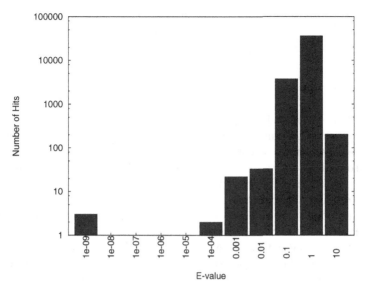

Fig. 2. Histogram of the E-values of sequences in the Test database declared significant by BLAST but not by $BLAST_{FDR}$

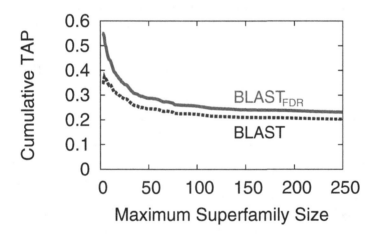

Fig. 3. Cumulative $BLAST_{FDR}$ TAP and BLAST TAP versus aggregate superfamily size for the Test database

were not retrieved by $BLAST_{FDR}$. For every dataset in the Test database, the retrieval list for $BLAST_{FDR}$ was shorter than the respective list for BLAST.

Furthermore, $BLAST_{FDR}$ performs notably better on datasets that belong to small superfamilies. Figure 3 illustrates this with the cumulative average TAP for both $BLAST_{FDR}$ and BLAST for ascending superfamily sizes. For example, for superfamilies with a size of twelve or fewer members, $BLAST_{FDR}$ has a TAP of 0.421 and BLAST a TAP of 0.332.

Similar results are obtained by using each of the ASTRAL40 database queries and searching in the NR database for up to five iterations and then using the resulting PSSM on the augmented database (data not shown).

4 Discussion

In this article we discussed an observed deficiency in the control of the proportion of irrelevant records in retrieval algorithms. Including too many irrelevant sequences has been shown to corrupt searches in a genetic database search algorithm [7]. To address this issue, we propose BLAST$_{FDR}$, an implementation of BLAST that exercises a false discovery rate method, for finer control over the percentage of irrelevant sequences.

Using accepted evaluation procedures, BLAST$_{FDR}$ had an average TAP value 14.1% higher than BLAST on the ASTRAL40 Test datasets. This difference is significant given the extremely wide use that BLAST enjoys. Furthermore, BLAST$_{FDR}$ is particularly appropriate for queries with small superfamily sizes as evidenced by it obtaining an average TAP value 26.8% higher than BLAST for superfamilies with sizes up to and including 12. For queries in larger superfamilies, if the goal is to assign function to a query, then adequately identifying the superfamily is sufficient. For example, retrieving 50% of a large superfamily clearly indicates which superfamily the query belongs. This objective is not currently captured in retrieval evaluation metrics and may make evaluation values misleading for large superfamilies.

While BLAST$_{FDR}$ does show significant performance improvements over BLAST, the increase was not seen for all queries. For example, Figure 1 illustrates that there are several datasets in the Test database that BLAST$_{FDR}$ receives a TAP value of 0.0 but BLAST achieves a non-zero TAP value. Clearly some improvements can be made to BLAST$_{FDR}$ to improve its performance.

Traditionally, the Receiver Operating Characteristic (ROC$_n$) [8] method has served as an evaluation criterion for retrieval efficacy. The ROC$_n$ method ignores the threshold implied by a homology search algorithm and truncates a list of matches after the nth irrelevant match. The resulting list of matches is plotted with the number of irrelevant matches on the x-axis and the proportion of relevant matches on the y-axis. A ROC$_n$ score is then the normalized area under the curve. Typically, $n = 50$. The ROC$_n$ method was not suitable for this study as it generally requires the threshold imposed by the algorithm to be artificially modified to allow for n irrelevant matches, thus erasing the affect of the threshold method.

While we used BLAST as an example in this study, other retrieval algorithms that use uniform thresholding could also benefit from the implementation of a FDR controlled threshold. Furthermore, employing more advanced false discovery rate methods, such as the Q-value method [12] could also yield improvements. Implementation of the Q-value, because it requires the entire distribution of statistical scores, is inherently challenging for a heuristic algorithm like BLAST.

References

1. Altschul, S., Gertz, E., Agarwala, R., Schäffer, A., Yu, Y.: PSI-BLAST pseudocounts and the minimum description length principle. Nucleic Acids Research 37(3), 815–824 (2009)
2. Altschul, S.F., Madden, T.L., Schäffer, A.A., Zhang, J., Zhang, Z., Miller, W., Lipman, D.J.: Gapped BLAST and PSI-BLAST: a new generation of protein database search programs. Nucleic Acids Research 25(17), 3389–3402 (1997)
3. Benjamini, Y., Hochberg, Y.: Controlling the False Discovery Rate: a Practical and Powerful Approach to Multiple Testing. Journal of the Royal Statistical Society, Series B 57, 289–300 (1995)
4. Bonferroni, C.E.: Il calcolo delle assicurazioni su gruppi di teste. Tipografia del Senato (1935)
5. Carroll, H.D., Kann, M.G., Sheetlin, S.L., Spouge, J.L.: Threshold Average Precision (TAP-k): A Measure of Retrieval Efficacy Designed for Bioinformatics. Bioinformatics 26(14), 1708–1713 (2010)
6. Chandonia, J., Hon, G., Walker, N., Lo Conte, L., Koehl, P., Levitt, M., Brenner, S.: The ASTRAL Compendium in 2004. Nucleic Acids Research 32(Database Issue), D189–D192 (2004)
7. Gonzalez, M., Pearson, W.: Homologous over-extension: a challenge for iterative similarity searches. Nucleic Acids Research 38(7), 2177–2189 (2010)
8. Gribskov, M., Robinson, N.: Use of receiver operating characteristic (ROC) analysis to evaluate sequence matching. Computers and Chemistry 20(1), 25–33 (1996)
9. Hochberg, Y.: A sharper Bonferroni procedure for multiple tests of significance. Biometrika 75(4), 800–802 (1988)
10. Holm, S.: A simple sequentially rejective multiple test procedure. Scandinavian Journal of Statistics, 65–70 (1979)
11. Hommel, G.: A stagewise rejective multiple test procedure based on a modified Bonferroni test. Biometrika 75(2), 383–386 (1988)
12. Storey, J.: A direct approach to false discovery rates. Journal of the Royal Statistical Society: Series B (Statistical Methodology) 64(3), 479–498 (2002)
13. Wheeler, T.J., Clements, J., Eddy, S.R., Hubley, R., Jones, T.A., Jurka, J., Smit, A.F., Finn, R.D.: Dfam: a database of repetitive DNA based on profile hidden Markov models. Nucleic Acids Research 41(D1), D70–D82 (2013)

A Pipeline to Characterize Virulence Factors in *Mycobacterium Massiliense* Genome

Guilherme Menegói[1], Tainá Raiol[1], João Victor de Araújo Oliveira[2],
Edans Flávius de Oliveira Sandes[2], Alba Cristina Magalhães Alves de Melo[2],
Andréa Queiroz Maranhão[1], Ildinete Silva-Pereira[1],
Anamélia Lorenzetti Bocca[1], Ana Paula Junqueira-Kipnis[3],
Maria Emília M.T. Walter[2], André Kipnis[3], and Marcelo de Macedo Brígido[1]

[1] Department of Cellular Biology, Institute of Biology, University of Brasilia,
70910-900, Brasília, DF, Brazil
[2] Department of Computer Science, Institute of Exact Sciences,
University of Brasilia, 70910-900, Brasília, DF, Brazil
[3] Department of Microbiology, Immunology, Parasitology, and Pathology,
Federal University of Goiás, 74605-050, Goiânia, GO, Brazil

Abstract. Virulence factors represent crucial molecular features for understanding pathogenic mechanisms. Here we describe a pipeline for *in silico* prediction of virulence factor genes in *Mycobacterium massiliense* genome that could be easily used in many other bacterial systems. Some few methods for this characterization are described in the literature, however these approaches are usually time-consuming and require information not always readily available. Using the proposed pipeline, the number and the accuracy of predicted ORF annotation were increased, and a broad identification of virulence factors could be achieved. Based on these results, we were able to construct a general pathogenic profile of *M. massiliense*. Furthermore, two important metabolic pathways, production of siderophores and bacterial secretion systems, both related to *M. massiliense*'s pathogenicity, were investigated.

Keywords: genome, rapid growing mycobacteria, pipeline, bioinformatics, nosocomial infections, virulence factors, metabolic pathways.

1 Introduction

1.1 *Mycobacterium Massiliense*

With the improvement of culture and identification techniques, the number of reported medical cases related to nontuberculous mycobacteria (NTM) has been greatly increased during the last few years [6]. Among those, mycobacteria of the *Mycobacterium chelonae-Mycobacterium abscessus* group composed of *M. chelonae*, *M. immunogenum* and, particularly, *M. abscessus*, arose as one of the most important opportunistic pathogens [14].

In 2004, Adékambi *et al.* [2] assigned a novel species for a closely related isolate, *Mycobacterium massiliense*, a representative species of rapidly growing

J.C. Setubal and N.F. Almeida (Eds.): BSB 2013, LNBI 8213, pp. 202–213, 2013.
© Springer International Publishing Switzerland 2013

mycobacteria (RGM). RGM have important implications in human diseases, as they are frequently associated to infections among immunocompromised patients as well as wound, skin, and soft tissue infections [14]. Additionally, these bacteria are naturally resistant to several classes of antibiotics, particularly to antituberculousis drugs. *M. massiliense* is characterized as a strictly aerobic, non-spore-forming, nonmotile, acid-fast, gram-positive rod that shares 100% of its 16S rRNA sequence with *Mycobacterium abscessus*. However, there is still intense debate in the scientific community whether or not *M. massiliense* should be considered a new species or simply a *M. abscessus* strain.

Ever since its description, *M. massiliense* has been increasingly reported as the responsible for soft tissue infection outbreaks. At the Midwest Region of Brazil, a major infection outbreak has been recently reported along with the association to antibiotic and disinfectants resistances that may have contributed to the difficulty in controlling the spread of this strain. In a previous work, our group sequenced and characterized the genome of a *M. massiliense* strain, which was isolated from wound samples of patients submitted to arthroscopic and laparoscopic interventions in Goiânia, Brazil [7]. This strain has been since then identified as "GO 06" and its complete genome is already available in GenBank [15].

1.2   Virulence Factors and Their Role in Pathogenesis

Some bacteria are known to be extremely virulent pathogens with the ability to cause infectious diseases, e.g. tuberculosis or salmonellosis. Pathogenic bacteria must be able to enter its host, to survive and to replicate inside the host cell, while avoiding the mechanisms of host cell protection. Therefore, bacteria present a set of molecular features in order to bypass or overcome the host defenses, which are commonly called virulence factors. Here we discuss two major virulence factor systems, which seem to play a decisive role in *M. massiliense*'s pathogenicity: the siderophores and bacterial secretion systems production pathways.

Siderophores are ferric ion specific chelating agents whose main role is to scavenge iron from the environment and make it available to the microbial cell. It is well known that the siderophore system is correlated to the virulence of some organisms, like *Yersinia enterocolitica* and *Erwinia chrysanthemi* [13], and there are evidences it has a function in *M. massiliense*'s pathogenicity, as ferric iron is an essential macronutrient for bacterial growth.

The pathogenicity of some bacteria, however, depend on their ability to secrete virulence factors, which can be displayed on the bacterial cell surface, secreted into the extracellular medium, or directly injected into a host cell. In Gram-negative bacteria, six systems have been described with this function, the bacterial secretion systems I-VI. However, recent studies have provided evidence that in Gram-positive bacteria, such as *M. massiliense*, an alternative protein-secretion system exists, the type VII secretion system (T7SS), which has five copies through the genome, named ESX-1 to ESX-5 [1].

1.3   A Pipeline for Virulence Factor Analysis

In this article, we propose a pipeline for virulence factor identification and annotation, which was used in *Mycobacterium massiliense*'s genome. This pipeline could also be applied to other bacteria, possibly increasing the available data on bacterial pathogenesis and supporting the development of counter strategies against pathogenic microorganisms.

Few *in silico* methods have been described to predict virulence factors in bacteria, most of them based on machine learning strategies, which consider common molecular features of known virulence factors to predict new ones [4,10]. These approaches rely on Support Vector Machine (SVM), a supervised learning strategy. For efficient characterization of virulence factors, these methods require a good quality input data, specially for the training phase, leading to time demanding programs. Furthermore, since virulence factors present a variety number of functions in the microbial cell, from a cell wall component to a secreted protein, finding common patterns to identify such molecules is really a challenging task.

There are also other methodologies based on phylogenetic information, which is difficult to get and therefore not always available [12]. In this context, we believe that our pipeline could be a simple yet very efficient alternative to characterize virulence factors in bacteria.

2   Methods

2.1   *M. Massiliense* and *Bacillus Anthracis* Genomic Data

M. massiliense GO 06 assembly data comprise a single chromosome, previously assembled by our group using MIRA [9], and two putative plasmids, named Plasmid I and Plasmid II, of roughly 60 and 96 kilobases, respectively. The ORFs (Open Reading Frames) were annotated with Genome Reverse Compiler (GRC), using a reference database composed of only mycobacteria genomes [16]. To validate the pipeline, all analysis were also done on the genome of a pathogenic bacteria from the Bacillus group, *Bacillus anthracis* str. Ames, whose sequences were downloaded from NCBI (AN:AE016879).

2.2   Pipeline Proposal

An overview of the proposed pipeline is shown in Figure 1. The pipeline is divided in four major steps: (1) Improvement of ORF annotation; (2) Identification of virulence factors; (3) Analysis of virulence factor genes in metabolic pathways; and (4) Construction of metabolic pathway maps. Steps 1 and 2 are automated by a single Perl script, as seen in Figure 2, which shows the management of the pipeline execution. The details of this pipeline are described below. All analyses were made on a desktop with an AMD Phenom II B95 X4 processor and 4 GiB of RAM, running Ubuntu 12.04.

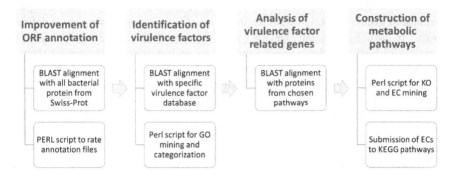

Fig. 1. Pipeline for identification and annotation of bacterial virulence factors

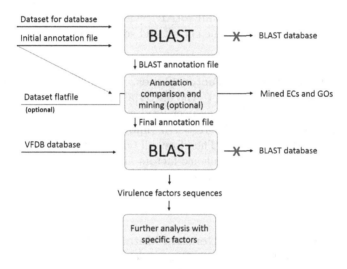

Fig. 2. Management of pipeline execution. The arrows leading to the boxes represent input files, while the arrows coming from them represent output files (those marked with a red 'x' are excluded by the end of the pipeline). All steps leading to the characterization of virulence factors are integrated in a single perl script.

Improvement of ORF Annotation. In order to have a well characterized sequence dataset, a BLAST [3] search with the initial annotation is performed, using a user-defined dataset for database construction. The E-value cut-off is also user-defined, having a default value of 1E-5.

The main part of the Perl script was designed to compare both the initial and the newly generated BLAST annotation files by assigning a score to each annotation and later choosing the highest score. For each BLAST query, the script verifies all the obtained hits, in order to choose the best annotation. The presence of keywords such as "hypothetical" or "putative" penalizes the score of a hit, while complementary features (e.g., Gene Name (GN), Enzyme Commission number (EC)) increases its score. This scoring method was defined in order to penalize uncharacterized sequences and to favour well annotated proteins, and both weights are user-defined upon initiation of the script. Initial tests with this pipeline suggested that penalizing keywords by 2 and adding 0.5 for complementary features avoided most bad annotation lines, therefore those weights were chosen as defaults.

The final gene annotation is the one with the highest score, obtained either from the initial file or from any of the BLAST hits. If many hits were equally good, the script chooses the one from the BLAST annotation file. In case of two or more BLAST hits with the same score, the one with lowest e-value is preferred.

The initial annotation of M. massiliense was provided by GRC, while the initial annotation file of B. anthracis was downloaded from NCBI. For both organisms, the dataset for the construction of the BLAST database was comprised of all curated bacterial proteins from UniProtKB/Swiss-Prot [5] (329,037 sequen-ces, as of April 2013). Both the E-value cut-off and scoring weights were the scripts' default.

Identification of Virulence Factor Genes. The annotation file generated in the last step is then compared to the bacterial protein database from VFDB (Virulence Factor Database) [8], a specialized repository of bacterial virulence factors. In this step, we are interested in finding genes related to virulence according to sequence similarity. Therefore, all ORFs with hits coming from the VFDB entries are considered virulence factors and stored in a new annotation file. The sequences filtered by this criterion had their GO (Gene Onthology) classification determined in the previous step, on an optional script module for managing flatfiles.

Analysis of Virulence Factor Genes in Metabolic Pathways. Considering that pathogenesis-related genes are often present in mycobacteria, we decided to analyse two of the most relevant metabolic pathways involved in this genus' pathogenicity: production of siderophores and bacterial secretion system proteins.

The gene sequences composing the chosen virulence pathways were downloaded from KEGG (Kyoto Encyclopedia of Genes and Genomes) [11]. By aligning the selected ORF sequences annotated as virulence factors with KEGG sequences, we could identify the genes related to the pathways and assign them a KO (KEGG Onthology). A Perl script was used in KO and EC mining, taking this information directly from the flatfile.

Construction of Metabolic Pathways. These results were used in the construction of metabolic maps for both pathways through KEGG, highlighting the genes present in *Mycobacterium massiliense* GO 06 genome. This analysis could not be performed for protein-secretion system VII (T7SS), since it is not yet fully described and there are no available corresponding pathway in KEGG.

3   Results and Discussion

The proposed pipeline allowed us to increase the number and accuracy of annotated ORFs. The assembled chromosome, which had initially 3,053 annotated ORFs, when compared to the final 3,388, showed an increase of 11% in ORF annotation. This also happened for both putative plasmids: Plasmid I initially had 33 annotated ORFs, which increased to 47 after being processed by this pipeline (42.4% increase), and Plasmid II, from 14 to 58 (314.3% increase) annotated ORFs. The data regarding the execution of the pipeline for both genomes can be found on Table 1.

Table 1. General data regarding the script execution. Vague annotations, such as "uncharacterized protein" were considered incomplete annotations. "I" means initial annotation; "W" weighted; and "U" unweighted.

	M. massiliense			*B. anthracis*		
	I	W	U	I	W	U
Incomplete annotations	1.659	997	997	2.503	1.923	2.088
Improved annotations	-	2.816	2.816	-	3.646	3.572
Elapsed time	-	1h28m19s	1h28m55s	-	2h26m31s	2h25m15s

Table 2. Distribution of virulence factor genes from *M. massiliense* GO 06 chromosome and two putative plasmids into GO categories

GO Category	Chromosome	Plasmid I	Plasmid II
Translation, ribosomal structure and biogenesis	5	0	0
Transcription	58	0	0
Replication, recombination and repair	5	8	0
Cell cycle control, cell division, chromosome partitioning	21	0	1
Pathogenesis	79	1	6
Signal transduction mechanisms	22	0	0
Cell wall/membrane/envelope biogenesis	9	0	0
Stress response	55	1	1
Posttranslational modification, protein turnover, chaperones	27	0	0
Energy production and conversion	14	0	0
Carbohydrate transport and metabolism	37	0	0
Amino acid transport and metabolism	49	0	0
Nucleotide transport and metabolism	7	0	0
Cofactor transport and metabolism	19	0	0
Lipid transport and metabolism	124	0	0
Inorganic ion transport and metabolism	46	0	1
Secondary metabolites biosynthesis, transport and catabolism	66	0	0
Antibiotic biosynthesis	25	0	0
General function prediction only	139	0	0

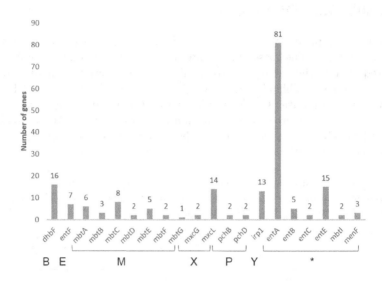

Fig. 3. Distribution of *M. massiliense* GO 06 genes related to siderophore production. The y axis represents the number of genes from a given family present in *M. Massiliense*'s genome. A letter code was assigned to each gene family, representing the siderophore production in which they are involved: Bacillibactin (B), Enterobactin (E), Mycobactin (M), Myxochelin (X), Pyochelin (P) and Yersiniabactin (Y). Gene families marked with '*' are either involved in the production of siderophore precursors or in more than one of the previously cited molecules.

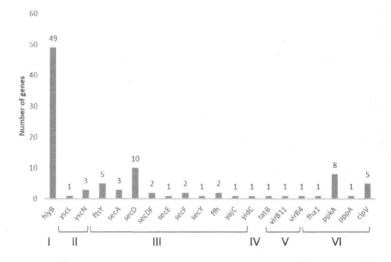

Fig. 4. Distribution of *M. massiliense* GO 06 genes related to bacterial secretion systems production. The y axis represents the number of genes from a given family present in *M. massiliense*'s genome. A number was assigned to each gene family, representing the secretion system production in which they are involved (I to VI).

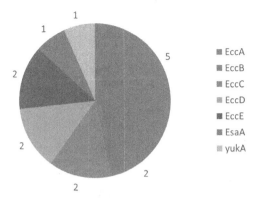

Fig. 5. Distribution of *M. massiliense* GO 06 genes related the Type VII secretion system production in *M. massiliense*'s genome

After identification and selection of all the virulence factor related genes, they were classified into a few selected GO categories for an easier overview of *M. massiliense*'s virulence profile. Table 2 shows this distribution for *M. massiliense*'s chromosome and both putative plasmids. Out of the 807 genes found to be related to virulence, 387 (48%) were genes involved in the organism's metabolism, most of them related to lipids (32%). In addition, 139 genes (17.8%) could not be properly characterized since their functions were poorly annotated (either they had no assigned GO or it was too unspecific, like "ATPase"). The high number of genes that could not be assigned to any class shows that there is still a limitation in protein databases related to virulence factors, even though most of the annotation came from curated sources.

Figures 3 and 4 show the number of detected genes in *M. massiliense* genome that compose the two studied metabolic pathways. Figures 6 and 7 depict the constructed metabolic pathway maps. Because of a severe limitation of KEGG pathways tool and database, some gene families were not correctly displayed and were highlighted manually, with an image editing software. This is the case of *mbt* genes, for example, which share the same EC number and could not be correctly displayed.

The majority of siderophore genes are from the *entA* family (EC 1.3.1.28, 81 copies), which might be explained by the fact that it is involved in the production of 2,3-Dihydroxybenzoate, a necessary precursor of the siderophores vibriobactin, enteroxelin, bacillibactin and myxochelin (Figures 3 and 6). This result indicates that the precursor could possibly be needed in a higher quantity in the bacterial cell. Other gene families involved in the production of siderophore precursors, such as *menF* (EC 5.4.4.2), *mbtI* (EC 4.2.99.21) and *entB* (EC 3.3.2.1), however, present a much lower number of genes in *M. massiliense* GO 06's genome. While most siderophore molecules seem to be produced by *M. massiliense*, yersiniabactin, vibriobactin and pyochelin are either not produced or have an alternative structure, as evidenced by the presence of only part of the gene families coding the precursor chemical structures.

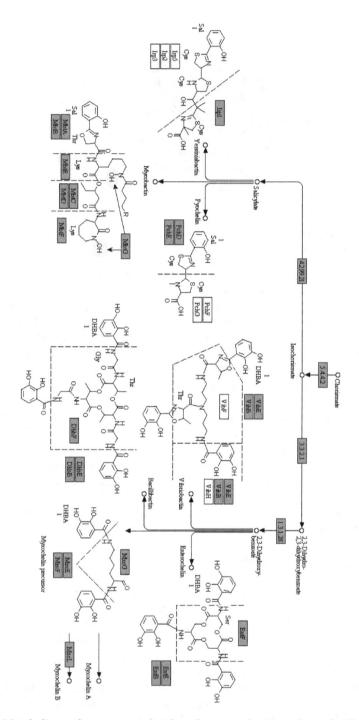

Fig. 6. Metabolic pathway map of siderophore production. Gene families in *M. massiliense* GO 06's genome are marked.

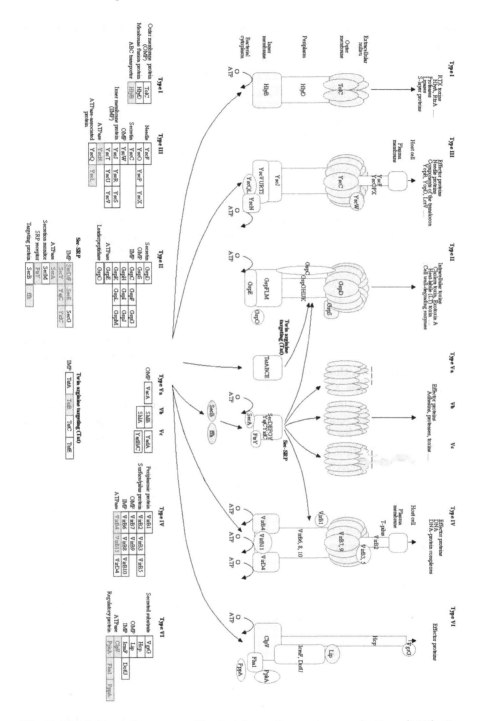

Fig. 7. Metabolic pathway map of bacterial secretion system productions (I-VI). Gene families in *M. massiliense* GO 06's genome are marked.

Genes from all secretion systems could be identified, most of them related to Type I (49 genes), followed by genes from Type III (27 genes). Additionally, 15 genes related to T7SS were found in *M. massiliense*'s genome, as shown in Figure 5, the majority of which belonged to the *EccA* gene family, whose products have ATPase activity, probably supplying energy for this system's functionality. However, none of the pathways were completely characterized, and key elements for the functionality of each system are still missing.

It is noteworthy that the characterization of virulence factors has never been done before for *M. massiliense* and the results obtained in the present study are valuable, mainly considering the relevance of this organism as a pathogen and as a study model to *M. tuberculosis*. By our sequence similarity approach, the greater the amount of information in virulence factor databases, the better the final quality of our characterization. This might have affected our analysis, specially for T7SS, a recently characterized system that does not have many related sequences available in databases. Therefore, additional data, such as transcriptome sequences, could be useful to improve this characterization.

4   Conclusion

In this work, we propose a pipeline to identify virulence factors, which can also be used to improve annotation in bacteria. Other approaches exist to solve this problem, e.g., methods based on Support Vector Machine, which are time demanding, and those based on phylogenetic information that are not always available. We believe that our pipeline is easy to implement and fast to produce refined results, when compared to the other tools. Even though we applied this pipeline to identify *M. massiliense*'s genes, it could be easily used in the characterization of other pathogenic bacteria, regardless their classification.

By analyzing *M. massiliense*'s virulence factors, we could define an overview of this organism's pathogenicity profile and verify the existence of important mycobacterial genes, which validate the consistency of the assembled genome. In addition, *in silico* data regarding isolate GO 06's genes related to siderophores and bacterial secretion system proteins were obtained, which could prove useful for developing strategies to control *M. massiliense* related outbreaks.

Acknowledgements. This work was supported by CNPq (grant numbers 301198/2009-8 and 564243/ 2010-8). G.M.R and M.E.M.T.W. were supported by research fellowships from CNPq, and T.R. by research fellowship from CAPES.

References

1. Abdallah, A., Gey van Pittius, N., Champion, P., Cox, J., Luirink, J., Vandenbroucke-Grauls, C., Appelmelk, B., Bitter, W.: Type "vii" secretion - mycobacteria show the way. Nature Reviews Microbiology 11, 883–891 (2007)

2. Adékambi, T., Reynaud-Gaubert, M., Greub, G., Gevaudan, M., La Scola, B., Raoult, D., MicGilbert: Amoebal coculture of *Mycobacterium massiliense* sp. nov. from the sputum of a patient with hemoptoic pneumonia. Journal of Clinical Microbiology 42, 5493–5501 (2004)
3. Altschul, S., Gish, W., Miller, W., Myers, E., Lipman, D.: Basic local alignment search tool. Journal of Molecular Biology 215(3), 403–410 (1990)
4. Andreatta, M., Nielsen, M., Aarestrup, F., Lund, O.: *In silico* prediction of human pathogenicity in the γ-proteobacteria. Plos One 5, e13680 (2010)
5. Bairoch, A., Boeckmann, B.: The "swiss-prot" protein sequence data bank. Nucleic Acids Research 20, 2019 (1992)
6. Carbonne, A., Brossier, F., Arnaud, I., Bougmiza, I., Caumes, E., Meningaud, J.P., Dubrou, S., Jarlier, V., Cambau, E., Astagneau, P.: Outbreak of nontuberculous mycobacterial subcutaneous infections related to multiple mesotherapy injections. Journal of Clinical Microbiology 47(6), 1961–1964 (2009)
7. Cardoso, A., Junqueira-Kipnis, A., Kipnis, A.: *In vitro* antimicrobial susceptibility of *Mycobacterium massiliense* recovered from wound samples of patients submitted to arthroscopic and laparoscopic surgeries. Minimally Invasive Surgery, 1–4 (2011)
8. Chen, L., Yang, J., Yu, J., Yao, Z., Sun, L., Shen, Y., Jin, Q.: "vfdb": a reference database for bacterial virulence factors. Nucleic Acids Research 33, D325–D328 (2005)
9. Chevreux, B., Wetter, T., Suhai, S.: Genome sequence assembly using trace signals and additional sequence information. In: Computer Science and Biology: Proceedings of the German Conference on Bioinformatics (GCB), vol. 99, pp. 45–56 (1999)
10. Garg, A., Gupta, D.: VirulentPred: a SVM based prediction method for virulent proteins in bacterial pathogens. BMC Bioinformatics 9, 62 (2008)
11. Kanehisa, M., Goto, S.: KEGG: Kyoto Encyclopedia of Genes and Genomes. Nucleic Acids Research 28, 27–30 (2000)
12. Nanni, L., Lumini, A.: An ensemble of support vector machines for predicting virulent proteins. Expert Systems with Applications 36(4), 7458–7462 (2009)
13. Neilands, J.: Siderophores: Structure and function of microbial iron transport compounds. The Journal of Biological Chemistry 270, 26723–26726 (1995)
14. Petrini, B.: *Mycobacterium abscessus*: an emerging rapid-growing potential pathogen. APMIS 114, 319–328 (2006)
15. Raiol, T., Ribeiro, G., Maranhão, A., Bocca, A., Silva-Pereira, I., Junqueira-Kipnis, A., Brigido, M., Kipnis, A.: Complete genome sequence of *Mycobacterium massiliense*. Journal of Bacteriology 194, 5455 (2012)
16. Warren, A., Setubal, J.: The genome reverse compiler: an explorative annotation tool. BMC Bioinformatics 10, 35 (2009)

Author Index

Akashi, Makoto   12
Alves, Ronnie   160
Amaral, João   181
Amman, Fabian   1
Araújo, Gilderlanio S.   104
Arruda, Wosley   136

Bernhart, Stephan H.   1
Bocca, Anamélia Lorenzetti   202
Bonnato, Diego   160
Braga, Marília D.V.   36
Brandner, Astrid F.   71
Braz, Fernando   181
Brígido, Marcelo de Macedo   136, 202

Campos, Sérgio   181
Carroll, Hyrum D.   194
Costa, Ivan G.   104
Cruz, Jader   181
Cunha, Luís Felipe I.   126

Dans, Pablo D.   71
Darré, Leonardo   71
Davis, Anthony G.   194
Dhalia, Rafael   94
Doose, Gero   1

Emmendorfer, Leonardo R.   148

Faria-Campos, Alessandra   181
Ferreira, Bruno   181
de Figueiredo, Celina M.H.   126
Figueroa, Nathaniel   170

Guimarães, Katia S.   24

Hausen, Rodrigo de A.   126
Hofacker, Ivo L.   1, 82
Hoksza, David   59
Höner zu Siederdissen, Christian   82

Imoto, Seiya   116

Junqueira-Kipnis, Ana Paula   202

Karro, John   170
Kipnis, André   202
Kowada, Luis Antonio B.   126
Krieger, Marco A.   94

Lins, Roberto D.   94
Liu, Xiaolin   170

Machado, Matías R.   71
Maranhão, Andréa Queiroz   202
Marques, Ernesto T.A.   94
Matsuno, Hiroshi   12
de Melo, Alba Cristina Magalhães Alves
   47, 202
Mendes, Marcus   160
Menegói, Guilherme   202
Miyano, Satoru   116

Niida, Atsushi   116

Oliveira, João Ricardo M.   104
Oliveira, João Victor de Araújo   202

Pantano, Sergio   71

Qin, Jing   1

Raiol, Tainá   136, 202
Ralha, Célia G.   136
Rosa, Rogério S.   24

Salvá, Thyago   148
Sandes, Edans Flávius de Oliveira   202
Silva-Pereira, Ildinete   202
Souza, Manuela R.B.   104
Spouge, John L.   194
Stadler, Peter F.   1, 82, 136
Stoye, Jens   36
Sundfeld, Daniel   47
Svozil, Daniel   59
Szépe, Peter   59

Tremmel, Georg   116

Viana, Isabelle F.T.   94

Walter, Maria Emília M.T.   136, 202
Wang, Jiajun   170
Werhli, Adriano V.   148
Will, Sebastian   1
Williams, Alex C.   194

Zeida, Ari   71